cx5359

The Asian Energy Factor
*Myths and Dilemmas of Energy,
Security and the Pacific Future*

Robert A. Manning

A Council on Foreign Relations Book

palgrave

The Council on Foreign Relations, Inc., a nonprofit, nonpartisan national organization founded in 1921, is dedicated to promoting understanding of international affairs through the free and civil exchange of ideas. The Council's members are dedicated to the belief that America's peace and prosperity are firmly linked to that of the world. From this flows the mission of the Council: to foster America's understanding of other nations—their peoples, cultures, histories, hopes, quarrels, and ambitions—and thus to serve our nation through study and debate, private and public.

From time to time, books and reports written by members of the Council's research staff or others are published as a "Council on Foreign Relations Book."

THE COUNCIL TAKES NO INSTITUTIONAL POSITION ON POLICY ISSUES AND HAS NO AFFILIATION WITH THE U.S. GOVERNMENT. ALL STATEMENTS OF FACT AND EXPRESSIONS OF OPINION CONTAINED IN ALL ITS PUBLICATIONS ARE THE SOLE RESPONSIBILITY OF THE AUTHOR OR AUTHORS.

THE ASIAN ENERGY FACTOR
Copyright © Robert A. Manning, 2000.
All rights reserved. No part of this book may be used or reproduced in any manner whatsoever without written permission except in the case of brief quotations embodied in critical articles or reviews.

First pubished 2000 by
PALGRAVE™
175 Fifth Avenue, New York, N.Y. 10010 and
Houndmills, Basingstoke, Hampshire, England RG21 6XS.
Companies and representatives throughout the world.

PALGRAVE™ is the new global publishing imprint of St. Martin's Press LLC Scholarly and Reference Division and Palgrave Publishers Ltd (formerly Macmillan Press Ltd).

ISBN 0-312-22437-0

Library of Congress Cataloging-in-Publication Data to be found at the Library of Congress.

Design by Westchester Book Composition

First edition: September 2000
10 9 8 7 6 5 4 3 2 1

Contents

Acknowledgments		v
Preface		vii
Chapter 1:	Introduction: The Apocalypse Industry	1
Chapter 2:	The Myth of Energy Scarcity: Energy Crisis or Cornucopia?	17
Chapter 3:	The Myth of Caspian Great Game and the "New Persian Gulf"	41
Chapter 4:	The Asian Energy Predicament: A New Geopolitics?	59
Chapter 5:	China: In Search of an Energy Policy	85
Chapter 6:	India: The Lumbering Elephant	119
Chapter 7:	Japan/Korea: Whither Asia's Mature Consumers?	143
Chapter 8:	Southeast Asia: Tigers as Regional Energy Link	169
Chapter 9:	Energy and Asian Security: Fueling Conflict or Cooperation?	187
Notes		209
Bibliography		235
Index		240

Acknowledgments

At the top of a long list of those to whom I must express my gratitude is Leslie H. Gelb and the Council on Foreign Relations, whose support and patience made this project possible. A special note of appreciation goes to Susan Tillou, whose research skills and diligence greatly facilitated preparation of the manuscript, and without whose administrative wherewithal the project would have floundered. I must also express my gratitude to Edward L. Morse and R. James Woolsey for their valuable input and for chairing the study group that helped shape this book. I am particularly indebted to Amy Myers Jaffe for her wisdom on the workings of the petroleum industry, insightful comments on the manuscript, and the invaluable series of energy studies of the James A. Baker III Institute for Public Policy of Rice University upon which I have liberally drawn.

This study also benefited from the contributions of Patrick Clawson, David Jhirad, Fereidun Fesharaki, James Clad and Dennis Eklof of Cambridge Energy Research Associates, Michael Lynch, Jon Choy, Michael May, Dennis O'Brien, Stephen Burns, and others who shared their work and insights with the study group. I would also like to thank Washington SyCip, a member of the Council's International Advisory Board, for his generous support of this project.

I am also very grateful to Zachary Davis for timely reality checks and helpful comments on several chapters, to Kevin Nealer and Eric Melby for their comments on the chapter on China, and to Kerry Rosenberg for his eleventh-hour help in preparation of the manuscript. I want to thank the many U.S., Japanese, and Chinese energy company officials and specialists (who shall remain nameless to protect the innocent); to apologize for any of the cast of thousands I have inadvertently failed to mention; and to underscore the point that I and I alone take full responsibility for what is in this book.

Some material from chapter 2 was incorporated in an essay published in the January–February 2000 issue of *Foreign Affairs* and some of the material on the Caspian in chapter 3 first appeared in *Survival,* Winter 1998–99. All charts, maps, and graphs used in this book have been reproduced with permission.

Last but not least, I reserve my deepest gratitude for my wife, Jan Austin, and my son, Aaron, whose indulgence and forbearance, in the end, made the arduous task of completing this project possible.

Preface

A bit like opening one door only to find it leads to another and yet another, this book grew out of a Council on Foreign Relations study group that explored the multifaceted challenges that the Asia-Pacific's mushrooming energy consumption poses to the world, to regional security, and how it is shaping the choices Asian face as they meet these challenges. There is little that is more essential to the daily life of civilization than energy. Access to stable, secure, and ample energy supplies is a fundamental underpinning of all nations' security and prosperity. In the modern era, the need to secure energy resources has been more than a minor consideration in imperial enterprises from Britain's expansion into the Near East in the latter part of the nineteenth century to Saddam Hussein's invasion of Kuwait in the last decade of the twentieth century.

Yet the cluster of issues surrounding energy questions do not fit neatly into simple classifications. The question of energy unavoidably connects otherwise disparate issues—economics, national security, and increasingly, the environment. However, these concerns tend to be perceived and addressed quite separately and independently of each other in terms of both intellectual categories and in the shaping of public policies. There is rarely any dialogue, let alone cross-fertilization, between specialists working on these distinct topics. Indeed, those focused on issues to traditional security, geopolitics, and military affairs seldom has a good grasp of the workings of world energy markets or in global environmental issues. Similarly, energy economists seldom focus on how regional security questions may have a profound impact approaching on energy flows. Like the proverbial blind men and the elephant, it is as if different specialists approaching energy-related questions from different perspectives arrive at very different assessments of the problem.

So in early 1998, when I began this project in earnest, what caught my attention was the intriguing fact that the driving force in global oil markets was the surging growth in Asian demand. For most of the 1990s (until the economic crisis interrupted the miracle economies of the Pacific Rim in late 1997) Asian demand grew at more than 5 percent annually. That is almost three times faster than that of the United States and Europe. And by most indicators, Asian demand will continue to grow at a faster pace than that of the rest of the world, accounting for perhaps half of the growth in global demand by the first quarter of the twenty-first century. Burgeoning Asian energy consumption and rapidly increasing exports raised a series of intriguing questions about implications for energy security, geopolitics, and even the environment.

It has been a dominant trend among those who study Asia-Pacific security to view energy as a negative factor in a regional situation characterized by widespread military modernization, an array of unresolved territorial disputes, lingering historical distrust, and suspicion looming behind strategic competition. Thus it became received wisdom, at seemingly endless conclaves on the future of Asian security, that the specter of competition for dwindling oil and gas resources was a likely source of future conflicts among increasingly well-armed states as they played out the region's panoply of territorial disputes in the twenty-first century. Asia's future, prominent political scientists suggested, may be Europe's past.[1]

But even as such prognostications became standard fare, oil prices were sinking to nearly $8 a barrel at the beginning of 1999—levels lower in real, inflation-adjusted dollar terms that at any time since the Great Depression. So low did prices drop that they generated new and unexpected collective discipline among producers, resulting in March 1999 OPEC accord that shut some 5 million barrels per day (b/pd) and pushed them back over the $30 range. Yet however volatile the market, few oil analysts project long-term price trends beyond current levels; few would be greatly surprised if prices drifted back down to single digits.

One delicious discrepancy that captured the disconnect between political scientists on the one hand and oil industry analysts on the other was the view of the South China Sea. In the oil business, by the 1990s, the term "South China Sea" became an insider joke: so many dry holes were hit that when a driller had to drill but expected to come up empty he often referred to it as "another South China Sea." Yet in most Asian security discussion the dispute over the Spratly Islands in the South China Sea, claimed by six countries, was frequently viewed as a potential source of major conflict in the Pacific in large part because the territory was—and is still believed by some—to be a treasure trove of oil and gas reserves!

Something was clearly wrong with this picture. It is precisely to try to bridge this gap in what I fear is a world of increasingly specialized, compartmentalized knowledge that I decided to pursue this book, to synthesize the economic and security dimensions of the energy question in Asia. My intent is to find the convergence of the economic and technical aspects of energy issues with the political and security aspects and elaborate the dynamics at play: the perils and the promise of the new geopolitics of energy; the policy choices for Asia-Pacific nations; and their consequences and policy implications for the United States.

As I explored these issues, I found what I began to view as series of myths and dilemmas. Demystifying the former and illuminating the latter serve as the organizing principles for my analysis. At the global level, there is the myth that the world is running out of oil, that massive price hikes and the end of the oil age is just over the horizon. Then there is the relatively new cottage industry that deems the newly independent states of Central Asia as the next "Persian Gulf," treasure trove of oil and gas resources, and the attendant twenty-first century version of the "Great Game" of strategic competition over control of these resources.

The first two myths are related to the third, the explosive growth in East Asian consumption, particularly by China and India, will spur new shortages and astronomical price hikes in oil and other resources. This in turn will foment more intense competition for scarce resources, and turn local territorial disputes into the major conflicts of the twenty-first century. All these elements of conventional wisdom are dubious, and what follows is an effort to demonstrate that this is the case, while offering alternative views of these phenomena.

At the same time, a host of intriguing yet underexamined questions arise from Asia's energy predicament. For example, as a result of the dynamic of global oil markets, East Asia is increasingly dependent on imports from the Persian Gulf/Middle East, while the United States imports decreasing amounts from that volatile region. These appear to be long-term trends. But what is their significance? Is the Gulf-Asian economic connection a harbinger of a new geopolitics of energy? What are the security implications for Asia and for the United States? Similarly, the beginnings of a new momentum toward environmentally friendly natural gas is evident throughout the Asia-Pacific. Does this trend open up new opportunities for cooperative development, for new patterns of interdependence? These are among the unanswered—in some cases, hitherto unasked—questions this book explores, along with some of the conventional wisdom it scrutinizes. I do not pretend to possess all the answers. Rather, my intention is to pursue further inquiry so that it is more likely that the right questions get asked and the wrong answers get questioned.

Chapter 1

Introduction: The Apocalypse Industry

Predictions are hard, especially about the future.

—Yogi Berra

One of the great things about being an intellectual in contemporary America is that there is rarely any price to be paid for being wrong. Unfortunately, there is too often little social virtue in being right, and in some instances—if one is too far from the received wisdom of the moment—there may be a punitive cost in disdain, vilification, or even ridicule. Western civilization of course will long be shamed by the terrible fates of Copernicus and Galileo, whose (then heretical) scientific truths about the cosmos were greeted by the blind wrath of the powers that be. On an admittedly more modest scale, reasoned, factual analysis forecasting a relatively bountiful future such as those made by economist Julian Simon and others languished in obscurity, often the object of scorn, while hysterical—and profoundly flawed—warnings of food and population crises were well received and grabbed the national limelight.[1] Indeed, those proclaiming imminent mass famine, fuel crises, and eco-catastrophe for a generation still capture our attention and shape the public policy agenda.[2]

While the focus of this book is on the many aspects of energy questions in Asia and their wider strategic significance, to be adequately understood, energy must be placed in the intellectual context of wider questions about humanity's relationship to the environment—issues of population, foods and other natural resources. These issues, playing out in the *zeitgeist* fostered by the onset of the environmental movement starting in the late 1960s, provide the political milieu in which thinking about oil and energy policies takes place. Obviously, population growth is the basic driving force behind neo-Malthusian doomsday predictions. It is no less a major determinant of energy needs and usage, and an equally important factor in energy use projections. And hydrocarbon use is a major environmental concern and hence an important factor shaping energy decisions, one that almost certainly will loom still larger in the next half century. Thus, viewing energy in this context is critical to fathoming the sense and nonsense of underlying assumptions and wildly wrong forecasts, and for distinguishing between real and illusory concerns in regard to energy.

Yet paradoxically the Information Revolution has rendered it more difficult to sort out myth and reality. The increasing laxity of news standards in the infotainment world of rapid-fire, staccato news cycles too often result in distortion rather than edifying truth. In the Information Age, sound-bite world at the end of the millennium, well-timed sensational prognostications—however noble the intent, and however accurate or inaccurate—tend to get attention, television airtime, foundation grants, and lucrative book contracts. This is true regardless of the author's sincerity or purity of motive. Moreover, this phenomenon also pervades areas far afield from those of resources, population, and environment, not least in international relations.

For example, the last decade of the twentieth century saw one candidate for the new post - Cold War foreign policy organizing principle after another rise and fall like shooting stars, while the proponents of such views prospered. For example, it seems now like ancient history, but the 1990s began with a torrent of books about Japan as the new global superpower. "The Coming War with Japan," screamed one. A superior "Japan model" was being emulated across a Tokyo-led Asia that would surpass the United States in a new "Pacific Century,"[3] another admonished. The endless volumes heralding the "Asian Miracle"—not least the World Bank's widely heralded 1993 book "The East Asian Miracle"—looked rather different after the regional economic malaise triggered by the July 1997 Thai currency crisis. More recently, "The Coming Conflict with China," as a much-debated 1997 book proclaimed, argued that an emerging Chinese behemoth as the key hegemonic challenge. Others warned of emerging global chaos, while sophisticated political scientists forecast a "Clash of Civilizations," as the West versus the Rest would become the twenty-first-century paradigm of foreign policy.[4] A cursory glance at East Asia, for example, where the principal risks of major war are essentially civil conflict among Confucian civilizations—North and South Korea, China-Taiwan—begins to cast doubt on such wisdom. One could find other illustrations of intra-civilizational rivalry if not conflict, such as Iraq's invasion of Kuwait.

But ephemeral post–Cold War intellectual trends pale before a far more basic concern: the specter of environmental disaster and resource shortfalls, not least of which is energy crises. This is by no means to cast aspersion on the pioneering efforts of many environmentalists who forced the issue front and center on the global agenda beginning in the 1960s and 1970s. Certainly, those sounding the clarion call of environmental alarm, beginning with Rachel Carson's classic *Silent Spring,* in 1962, dramatized what indeed proved to be profoundly serious problems.[5] Unquestionably, such alarmist views helped create a new sensibility. They generated a critical mass of popular awareness of the environment and collective anxiety that helped change public and private sector behavior. Indeed, it can be argued that the environmental movement ushered in a generational shift from the post–World War II generation in perceptions of nature: from a view of nature as something to be conquered or tamed to the baby boomers, and following generations' worldview of nature as something requiring careful stewardship, something with which one should strive to be in harmony. Focusing high-profile attention on hitherto largely ignored problems unquestionably forced society to address the foul air, polluted water, and other environmental degradation.

However, there has been a downside. A snowballing environmental movement, however, produced tracts admonishing, as Paul Ehrlich did in 1968, for instance, that population growth would result in mass starvation not only in an overcrowded Third World but in the United States as well. Things turned out rather differently. Yet a generation later, in 1992 in his book *Earth in the Balance,* then Senator Al Gore echoed the timeworn Malthusian "population crisis" argument. Over the past generation, warnings of dire crisis have on more than rare occasions proved wrong or greatly exaggerated. Yet a combination of suspicion of government and, to varying degrees, popular distrust of multinational corporations—amplified by a sensationalistic media—fostered a still lingering popular psychology of worst-case expectations, a culture of alarmism, if not conspiracy and presumed malevolence. At the extreme end, "deep ecologists" with an absurdly idealized view of the natural world argue it is the human race itself that is a virus unto nature.[6]

Of late, the experience of the past three decades has begun to swing the pendulum back to a more balanced, mainstream view of environmental problems in the realm of public policy, one where being "green" can be not only compatible with, but advantageous for business, and where costs and benefits tend to be weighed and debated in public policy discourse. Yet it sometimes seems that almost any charge of malfeasance or imminent danger tends to be believed as true unless proven false, rather than the burden of proof being placed on those making the assertions. Witness the "alar" apple scare in the 1980s, or allegations in the mid-1990s that electric power lines are a cause of cancer. Good news is most often downplayed if not unwelcome.

One recent example of this phenomenon may be the unproven public health scare that synthetic chemicals (such as DDT and dieldrin) act as "endocrine disrupters" in humans, claims growing out of a well-promoted, best-selling 1996 book, *Our Stolen Future*. In an August 30, 1999, *New Republic* critique, Gregg Easterbrook pointed out that when the National Research Council, part of the National Academy of Sciences, declared in August 1999 that the theory was "rife with uncertainties" and unsupported by health data it was largely ignored. While the theory may prove to be true, the Academy called for more research to be conducted before one could arrive at any such judgment.

Nonetheless, the idea had become a hot topic on talk shows, in the Congress, and no less than Vice President Al Gore cited "a large and growing body of scientific evidence" indicating a threat. Easterbrook analyzed dubious studies and oft-cited scientific papers subsequently withdrawn quietly to demonstrate claims hyped despite flimsy evidence. After cataloging a long list of ecological problems showing positive trends from the banning of ozone-depleting CFCs to the return of endangered species like the bald eagle and falcon, Easterbrook (author of an important work, *This Moment on Earth,* making the case for a more balanced second generation approach to the environment) concludes, "The politics of instant doomsday require something fresh and suitably bleak."

The fear of humans inflicting irreversible damage to the earth extends to the notion of natural resources being finite. Both ideas have long underscored the fragility of civilization—and has lent credibility and moral force to a seemingly interminable flow of doom-and-gloom forecasts about the fate of a planet with ever increasing needs. Energy, providing light, heat (and refrigeration), and not least, transport, is the essential lifeblood of industrial and postindustrial life. Moreover, energy is an enabling resource, making it possible to exploit, if not create, other resources.

Oil has been ascendant as the most widely utilized and internationally traded source of energy throughout the twentieth century, certainly since Winston Churchill made the fateful decision in 1913 to shift the power source of the British navy from coal to oil and the mass production of automobiles began to take off. Indeed, a stable and secure supply of oil is and will remain for much if not all of the twenty-first century a keystone of security for all nations.

It was no mere coincidence that Hitler sought the oil fields of Baku when Germany invaded the Soviet Union, or that Japan grabbed the oil fields of northern Sumatra in Indonesia early in its rush to empire in the 1930s. Indeed, Japan's preemptive strike at Pearl Harbor was in no small measure prompted by a U.S. oil embargo. The concentration of more than two-thirds of the world's oil reserves in the unstable region of Southwest Asia is and will remain a basic geostrategic fact shaping international relations well into the twenty-first century.

Like many of the Earth's treasures, oil appears a finite resource. Thus, it is not surprising that even as the price of oil—not to mention most minerals from aluminum to zinc—approached record lows in the late 1990s, a steady stream of missives in the popular and elite media, policy journals, and even distinguished scientific publications cautioned against the mood of complacency that has set in since the days of gas lines and skyrocketing prices of the 1970s.[7] The days of cheap, plentiful fossil fuel, they admonish, are numbered.

Indeed, the energy question is but one aspect of the well-received litany of prophesied catastrophes generated by humankind's ever-growing demands on the limited bounty of the earth. Not only do new energy crises replete with shortages, gas lines, and price hikes loom, it is argued, but as one prominent and usually levelheaded foreign policy analyst earnestly observed in 1996, "The closing years of the twentieth century are likely to be known as an era of resource scarcity."[8] It is a now familiar apocalyptic scenario: world population continues to mushroom to increasingly unsustainable levels, we are warned. Food production cannot keep pace, triggering mass starvation; minerals will be exhausted; fossil fuels wreak havoc on the environment; eco-catastrophe looms just over the horizon.

There is, of course, nothing novel about such alarmism. In his book *How Many People Can the Earth Support?* Joel Cohen cites ancient Babylonian writings and Chinese scrolls more than two millennia ago warning of overpopulation.[9] The contemporary debate, of course, has its origins in the waning years of the eighteenth century. The Enlightenment thinker Marquis de Condorcet argued that the powers of human ingenuity and reason could conceive the technology to support an expanding population, and that a more educated populace would lead to falling birth rates.

While the fullness of time has revealed that Condorcet has thus far clearly prevailed in the debate, he is long forgotten. His contemporary interlocutor, however, who was inspired to respond to Condorcet's arguments, is, of course, a veritable household name, synonymous with environmental alarmism: the English cleric Thomas Robert Malthus. Though egregiously wrong, it is Malthus whose work, including the famous 1798 "Essay on the Principle of Population," lies at the root of modern environmentalism. Its alarmism and simple logic proved more compelling. Malthus predicted with seemingly impeccable reasoning that since population growth was geometric, while food supplies increased only arithmetically, starvation in Britain was an impending reality. Thus did doomsaying become a mainstay of the post-Enlightenment Western intellectual landscape.

Much current angst is exacerbated by a widespread expectation that the center of gravity of world economic activity—and increasingly, military capacity—is shifting from the Atlantic to the Pacific Rim. In 1960, East Asia accounted for just 4 percent of the global GNP, by 1992, that grew to 25 percent of the world economy. The Asia-Pacific, with over 50 percent of the world's population, has surpassed Europe in its share of the world's military spending, in consumption of energy, and collectively holds nearly one trillion dollars in foreign reserves. The reemergence of China, and more broadly the rise of Asia after a generation of unprecedented economic growth, has given new impetus to those fearful that the emerging middle classes of the Pacific Rim would produce the ecological catastrophes and resource depletion about which environmental Cassandras have been warning for three decades.

It does not require great imagination to draw calamitous ecological conclusions about a world in which, say, China and India mass-produce "people's cars," and in 2020 there are an additional 400 million gas-guzzling autos on the roads of the two Asian giants—nearly equal to the total number of cars worldwide in 1996. Or consider the potential impact on food prices and supply if a new middle class of similar magnitude was eating meat several times a week instead of primarily rice and vegetables. Between 1978, when its economic reforms were launched, and 1994 China's consumption of pork quadrupled to some 30 million tons, requiring some 100 million tons of grain to produce.

The Club of Doom

Perhaps it is a natural human impulse to be anxious and fearful of what the future holds; to instinctively give credence to dark projections about what lies just over the horizon. Yet for nearly three decades in soon-to-be forgotten best-seller after best-seller pessimistic "experts" have been wrongly predicting famines, resource depletion, and global catastrophe. One of the most celebrated Malthusian cosmic screams in the post–World War II era was that of Paul Ehrlich, who ominously warned in a 1968 best-seller, *The Population Bomb,* "The battle to feed all of humanity is over. In the 1970s the world will undergo famines—hundreds of millions of people are going to starve to death in spite of any crash programs embarked upon now." The book sold well over a million copies, helped shape the environmental movement that took off in the 1970s, forged a new awareness and sensitivity about the environment and a mind-set of scarcity and impending apocalypse—one all but impervious to unwelcome facts. Indeed, a generation later in the 1990s, Ehrlich was—with no less sense of urgency—still publishing books about the population explosion. Others sharing his perspective were passionately warning of impending environmental degradation and food shortages that would stem from "overpopulation."[10] The intellectual archaeology of the 1960s and 1970s environmental books genre reveals—with little digging—no shortage of such popular writings by engaged experts with titles like *Famine 1975.*

Paul Ehrlich then, was hardly a lone voice but rather something of a trend-setter. Indeed, he was at the forefront of what fast became a widespread view of humanity's fate. Since 1973, Lester Brown of the Worldwatch Institute has been regularly warning of impending food crises. This warning that disaster for humankind loomed just over the horizon was dramatically portrayed and greatly amplified by a highly influential 1972 report entitled, *Limits to Growth,* by a group of prominent scientists and public notables called the "Club of Rome." The report, published as a book, sold over 4 million copies in more than 20 languages and became a veritable new paradigm for thinking about the relationship of the economy to the environment. "Zero growth" became a rallying cry. If the human race did not self-impose limits, natural limits in the form of starvation, exhausted resources, and an environmentally degraded planet would impose them at our peril.

The Club of Rome suggested that the world could run out of oil reserves by the 1990s and that reserves of aluminum, copper, uranium, natural gas, lead, and zinc would approach exhaustion in the not too distant future. "If present trends in world population, industrialization, pollution, food production and resource depletion continue unchanged," the report asserted, "the limits to growth on this planet will be reached sometime within the next one hundred years. The most probable result will be a rather sudden and uncontrollable decline in both population and industrial capacity."

The Club of Rome forecast was based on what is called a "system dynamics" model originally developed by an MIT professor, Dr. Jay W. Forrester. The social science construction drew conclusions based on extrapolations of trends in the interaction of population, agricultural and industrial production, natural resources, and environmental degradation, premised on exponential growth. It sought to represent real world phenomenon. But so controversial was the effort in the scientific community that it inspired a critique a year later by a group of scientists at Sussex University in the United Kingdom who released a book called *Models of Doom*.[11] The Sussex group made a persuasive case that the assumptions of the modelers as well as the data were seriously flawed. In a variation of the computer adage "Garbage in, garbage out," they suggested that "Malthus in, Mathus out" might be a fair assessment of the degree to which the questionable assumptions of the modelers shaped a flawed outcome.

One key point is that some of the data is unknown or unknowable. World mineral reserves, for examples, may not be infinite, but how much the earth contains and what portion might be commercially extractable with what technology is not among the available data. Nor is it easily anticipated. Over the half century from 1940 to 1990, estimates of known reserves were constantly revised upward, for a host of minerals: lead, zinc, copper. Known reserves of bauxite grew by an average of 50 million tons a year between 1941 and 1953; from 1950 to 1958 they grew by 250 million tons a year.[12] One of the *Limits to Growth* modeling scientists estimated that world bauxite reserves would be exhausted in forty-nine years. The list goes on. But clearly the Club of Rome misunderstood the concept of known reserves—that which is commercially exploitable at current prices with current technology—as an indicator of ultimate recoverable reserves. The latter is something that is simply less knowable, but certainly declining prices are not an indicator of scarcity.

Similarly, the degree to which the behavior and values—let alone public policies—of human beings may change is not something that is likely to be modeled with any semblance of accuracy. Consider, for instance, the remarkable shifts in public and corporate attitudes and government policies toward conservation, or toward tobacco—not to mention innovations in clean technology—over the past generation that have qualitatively improved the environment. Automobiles today produce roughly 95 percent less pollution than those built before 1970, in part due to regulatory changes forced by the 1970 Clean Air Act (e.g., curbing tailpipe emissions, the introduction of the catalytic converter, etc.). In sharp contrast are the disastrous economic and environmental consequences of the policies pursued by communist regimes in the former Soviet Union and Eastern Europe. The Club of Rome's own website concedes that its model "could

Introduction: The Apocalypse Industry   9

not account for political factors, psychology, technical progress or unforeseeable continuities." Such a remarkable admission might lead one to wonder what the point of the exercise was in the first place—and why anyone took it seriously. Yet undeterred, the group actually released a book in 1995, *Beyond The Limits,* which said that although its previous view of the future was far too dark, it was equally pessimistic now about the future![13]

Perhaps the single most dramatic Club of Rome shortcoming, one that the Sussex group identified as being "at the heart of our differences," is technological change. The Information Age technological revolution in the oil industry, as discussed in the following chapter, is a case in point. But it is worth recalling that for thousands of years, until less than four centuries ago, coal was not considered a prime energy source. More telling, just a century ago, few people thought that oil and autos would be hugely important, much less come to veritably define energy and transport in the twentieth century. Similarly, commercialization of fuel cell technology or breakthroughs in areas like fusion technology—which could provide an infinite source of energy—or other new technologies (discussed in the next chapter) are not fully predictable but nonetheless certainly possible, even likely. Indeed, if past is prologue, developments that could radically transform the energy situation in coming decades are probable.

In the area of food production, for example, high-yield techniques and other improvements produced remarkable results: in the late 1930s, U.S. production of 17 essential food and fiber products (wheat, corn, cotton, etc.) was 210 million tons from 77 million acres. Five decades later, in the late 1980s, production of the same commodities rose to 600 million metric tons from 72 million acres.[14] Yet it is difficult to imagine sufficient data to allow inputs for modeling that can anticipate potential breakthroughs in technology. One could go on detailing the flaws of the now oft-maligned Club of Rome future simulation exercise—and it was roundly criticized at the time, not only by the Sussex group—but by a broad range of economists.

Despite its demonstrable defects, so compelling was the message that as Julian Simon wrote two decades later, *Limits to Growth* continues to be cited in the popular press as authoritative." A 1974 CIA report that was leaked to the press, *Potential Implications of Trends in World Population, Food Production and Climate,* is another interesting reflection of the prevalent scarcity mind-set of the time. The report concluded, "Trying to provide adequate world food will become a problem of overriding priority in the years and decades immediately ahead." Interestingly, in light of the current debate on global warming, much of the report's pessimism was based on the fear of an imminent global *cooling* trend, which it said reflected a "near consensus" among climatologists." This led to the conclusion that "there would almost certainly be an absolute shortage of food."[15]

Indeed, a measure of the Club of Rome influence was evidenced in 1980, when the Carter administration released the results of a major exercise called *The Global 2000 Report to the President,* a massive three-volume, graphs-and-statistics-laden tome. Its particular significance was that the leader of the free world had elevated the discourse to the official level. Unlike the Club of Rome's report, *Global 2000* qualified its conclusions with a host of caveats about the limits of its modeling methodology. The study limited the reliability of its projections to "directional indicators." Nonetheless, much of its prognosis was at great variance with the reality that unfolded two decades later. The projected population at century's end of 6.35 billion is off by nearly 400 million, but the *Global 2000* suggestion that by the end of the twenty-first century it "would approach 30 billion" will likely prove to be off by an enormous order of magnitude. Demographers are still uncertain about precisely what causes fertility rates to decline. Nonetheless, current median-fertility UN projections (discussed below) are for just over 10 billion at the end of the new century, while several equally plausible scenarios suggest a decline in world population beginning in midcentury.

The Carter administration study was wide of the mark in regard to the fate of natural resources, projecting that "many of the most vital resources are projected to rise in real terms." For example, *Global 2000* explained, "To keep energy demand in line with anticipated supplies, the real price of energy is assumed to rise more than 150 percent over the 1975 to 2000 period." The report was more sanguine in regard to nonfuel minerals, predicting that none would be approaching exhaustion. In regard to the array of environmental questions linked to food and agriculture, air and water pollution, and rain forests, the report's projections were closer to indicating the broad direction of trends than to making precise numerical forecasts. Interestingly, on the issue of global climate change, *Global 2000* suggested three possible scenarios—status quo, global cooling, and global warming—to which it assigned varying probabilities. The dominant conclusion was uncertainty about the future. In sum, in broad terms, *Global 2000,* with more nuance, detail, and sophistication, echoed the Club of Rome theme of a planet increasingly in grave peril.

Happy Mistakes

What is truly remarkable is that such prognostications not only have been wrong *but many now appear exactly backwards!* This is the case even if one posits that the dire warnings sparking corrective action were one of the factors leading to more benign outcomes.

The world's population has indeed mushroomed from 1.6 billion in 1900 to 3.5 billion in 1970, to some 6 billion by 2000, and in light of the concentration of population growth in poor countries, one need not question the wisdom of population control efforts or the urgency in narrowing the gap between rich and poor countries. Yet rather than hopelessness and famine, by almost every measure the human condition is significantly improved in relative terms, even for the poorest. Despite dreadful disparities both between and within nations, in historic terms, on average, people now live longer, are healthier, consume more, and are more productive than in the past. Mortality rates are down and continue on that trajectory, even in less developed Third World nations like Bangladesh, by 40 percent or more over the past two generations, and food production has expanded more than 20 percent faster than population. Given that the past century has seen the fastest expansion of human population in ten millennia, this achievement appears all the more remarkable. The price of not only food but virtually every mineral has dropped steadily over the long term in real dollars, while reserves of most actually grew. Oil prices have been at historic lows and proven reserves at unprecedented high levels. While there are new environmental problems on the agenda, in OECD countries, many of the first-generation environmental problems—polluted air, rivers, and lakes, deforestation, etc.—have improved markedly, though much of the developing world has been slow to address such problems. There is 30 percent more forest today in Europe than there was half a century ago. One hundred fifty years ago, Vermont, Massachusetts and Connecticut were 35 percent forest; today they are 59 percent forest, and the United States on the whole has more forested areas now than it did in 1920![16]

The vast majority of Chicken Little admonitions about the plight of the human race and the resources of the earth over the past generation, to put it mildly, do not correspond to current realities. This substantial gap suggests that a better understanding of this phenomenon and the basic assumptions that led to such flawed forecasts may be instructive in gauging energy futures. In particular, the sea change in forecasts about population growth, which has been viewed as a basic causal factor in regard to food and natural resource shortages (e.g., more people, more pollution, faster exhaustion of resources), merits a closer look. After all, there is now serious concern about an impending population *implosion* beginning barely two generations into the twenty-first century, an outcome with far-reaching consequences for energy security.

12   The Asian Energy Factor

Despite enormous global population growth in the twentieth century—it more than tripled—this increasingly appears deceptive as any indicator of long-term trends. To project such trends indefinitely would be a bit like saying that since a baby doubled or tripled in size its first two years it would continue to do so indefinitely. Historically, demographic changes have not been smooth and consistent. As one demographer humbly conceded, "there is no accurate method for forecasting the future rate of population growth for a country or a region, much less the world."[17] A 1969 U.S. State Department forecast, based on UN assessments, was for 7.5 billion by 2000. By the mid-1970s, the UN Fund for Population Activities was predicting "nearly seven billion." At the other end, in 1977, the often alarmist Worldwatch Institute dropped its estimate to 5.4 billion by 2000, a number reached in 1990.[18] Clearly, demographic forecasts are at best a rather inexact science.

Over the past four centuries, there have been large increases and declines. Over the span of human existence, there have been spurts, from the time of tool-making perhaps a million years ago, roughly 10,000 years ago when agricultural production and animal herding led away from hunting and

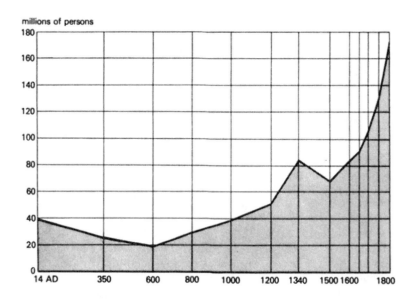

Figure 1.1 The Population of Europe, A.D. 14–A.D. 1800
Source: Julian Simon, "The Ultimate Resource 2," Copyright © 1996 by Princeton University Press. Reprinted by Permission of Princeton University Press.

gathering to a more sedentary existence. In both cases, after the initial productivity gains from the new technologies played out, population that had grown rapidly began to level off.[19] (See Figure 1.1.)

One major factor in the world population increase over the past two centuries has been the decline of the death rate. Changes produced by science and technology in medicine, sanitation, and the like played a large role in qualitative leaps in life expectancy from birth. Over the millennia, life expectancy improved only slowly and incrementally, from age 20 to 30. But in the past two centuries, and in less developed countries, especially since World War II, it has jumped enormously. (See Figure 1.2.)

Another trend increasingly evident over the past two centuries is a consistent pattern of steady decline in fertility rates. This is often attributed to shifting attitudes and behavior associated with increasing rural-to-urban migration and more broadly to "modernization," but there is no firm consensus among demographers on the reasons for it. Nonetheless, this trend has been especially pronounced in the twentieth century. Total fertility rate (TFR), the average number of births per woman, in the last half of the twentieth century dropped dramatically, from 5 to 3.1 children per woman

STANDING ROOM ONLY?

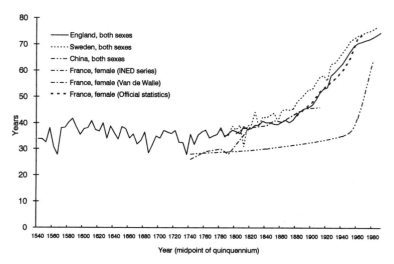

Figure 1.2 Life Expectancy in England, France, Sweden, and China
Source: Julian Simon, "The Ultimate Resource 2," Copyright © 1996 by Princeton University Press. Reprinted by Permission of Princeton University Press.

between 1950 and 1990. In the industrialized nations, the TFR fell from 2.8 to 1.7; in less developed areas, it was even more pronounced, from 6.2 to 3.5. Certainly, contraceptive technologies and improved infant mortality resulting from things like better health care and improved sanitation have both been major factors. In a number of countries such as Sweden, Italy, and Japan, fertility has been below replacement rates for several decades.

These changes have begun to alter mainstream thinking about population in what may amount to a paradigm shift over the coming decade. Even as chronic Cassandras still beat the "population explosion" drum, there is a growing scientific consensus that a population implosion is highly probable beginning by the mid-twenty-first century. As one demographic analyst wrote, "The evidence now indicates that within fifty years or so world population will peak at about eight billion before starting a rapid decline."[20] This dramatic shift among the world's leading demographers is reflected in the most comprehensive effort to project future trends, the United Nations "World Population Prospects." Significantly, UN projections continue to be revised—downwards. Beginning in the late 1990s, several of its seven scenarios—none of which are implausible, and all of which are indeed in the median range of the spectrum of possibilities—anticipate zero growth by midcentury and a declining world population thereafter.

The low and low/median scenarios in the 1998 UN population projection, both based on the very likely continuation of modestly lower fertility trends, forecast stabilization by 2050 at between 7.7 to 8 billion gradually declining over the subsequent hundred years, to 6.4 in the low/medium and 3.6 billion in the low scenario by 2150. Such remarkable possibilities are by no means based on extreme assumptions. Even under the median scenario, the projection is for 8.9 billion in 2050. Indeed, the debate, seems to be over whether world population will be closer to 8 billion than 11 billion before it levels off. The UN Population Conference's so-called Cairo Agenda is for a $17 billion program aimed at limiting world population to 9.8 billion by 2050.

Of course, the demographics of decreased fertility are not uniform. Indeed, they indicate a significant shift toward much larger concentrations of population in Asia, Africa, and Latin America (which already hold over half the world's population, increasing to about two-thirds by 2025). This population shift also suggests that the "North-South" gap between the developed and developing nations may also grow. The trends also show a broad "graying" trend in populations, more pronounced in OECD countries, altering the age distribution in many countries with myriad policy implications (e.g., labor force, health care, pensions, etc.).[21] Nonetheless, this suggests a very different set of problems than the neo-Malthusian Cassandras fear of the perils of scarcity on an overcrowded planet bursting at the seams.

Still more off the mark than population forecasts have been predictions about the purported consequences of what has indeed been rapid population growth—more than doubling in the last half of the twentieth century: famine, skyrocketing food prices, resource exhaustion. As noted above, the prices of most minerals have steadily declined. And what about food? No less a doomsayer than Lester Brown wrote in his annual *State of the World Report* for 1998, "Between 1950 and 1993, world prices of wheat, corn, and rice fell in real terms by 67, 83 and 88 percent, respectively." As he has done over the past two decades when prices rise, Brown argued that grain price rises in the mid-1990s were "the first economic indicator that the world is on a path that is environmentally unsustainable." Yet the drop in prices—the *Economist* food index is down 43 percent from 1997 to fall 1999—and U.S. farm crisis at the end of the decade suggest once again that short-term fluctuations have been mistaken for wider trends.

The fact is that per capita food production has steadily increased in the twentieth century more than 20 percent faster than population. More food is increasingly grown on less land at lower prices with modern agricultural techniques, biotechnology, and other technological advances shattering Malthusian predications. According to the UN Food and Agricultural Organization (FAO), per capita global calorie availability rose by some 30 percent. For the less developed regions, it rose close to 40 percent.[22] Between the early 1960s and late 1980s, according to the FAO, per capita food supplies rose by 18 percent worldwide, and about 28 percent in less developed regions.

As one demographer summed up, "A dramatic improvement in the world's food situation has occurred. it has taken place steadily over consecutive generations, and this period of sustained improvements coincided with the most rapid episode of global population growth in human history."[23] This is not to ignore or downplay the problems of crushing poverty and uneven distribution of wealth among and within nations. But it should be noted that some of the worst famines—the Ukraine under Stalin in the 1930s, China during the Great Leap Forward of the 1950s, Ethiopia in the 1980s, and North Korea in the 1990s—have been the result of political and/or policy factors other than an overburdened population. Most frequently they are the result of political factors: misrule, fatally flawed communist economic policies, civil wars, and in some cases, as seen in Sudan, famine and food distribution as a political weapon. Nobel Prize economist

Amartya Sen argues that there have been no famines in democracies.[24] That such population changes can be the consequence of wrongheaded policies is also evident in the regressive mortality rate in Russia after the demise of the Soviet Union.

Economist Julian Simon sought to explain one reason for flaws in forecasting scarcity by drawing a distinction between different methods—the engineering "technical method" and that of the economist. The technical method would estimate the known quantity of the resource, extrapolate the future use rate from the current use rate, and subtract the sum total from the known "inventory." The economist's approach is based on price trends. Is there reason to believe the future will be different than the present? If the data is clear and consistent—such as decades of declining mineral and food prices—then one can project the trend into the future. This is particularly true in the case of "known reserves." An economist would consider this category of resources as current inventory, something to be replenished when it begins to be reduced, rather than as something finite, total stocks that are being depleted. Thus far, the economist's method appears by all measures closer to the truth, with the caveat that forecasting remains as much art as science. But innovative efforts and technologies have resulted in reserves of most minerals increasing over time, while prices have tended to drop.

This brief survey of forecasting trends in population, environment, and resources, would above all, counsel humility in the exercise of prognostication. Yet as we have seen, there are clearly salient patterns that should inform any exploration of the consequences of Asia's burgeoning energy needs. One should never discount the Law of Unintended Consequences. Nonetheless, the patterns of price and availability in minerals and other resources suggest that in the realm of energy, the Asian factor—and what we do not know—may not have the draconian impact on the future of fossil fuels, the environment, or regional security that many analysts fear.

Chapter 2

The Myth of Energy Scarcity: Energy Crisis or Cornucopia?

Some 75 million barrels of oil a day (b/pd) were consumed around the world in 1999. The International Energy Agency (IEA), perhaps the most authoritative source of oil data, projects that world demand is likely to grow to roughly 110 million barrels a day by 2020.[1] The Asian factor will be a driving force shaping the world energy situation: nearly 50 percent of that growth in demand, about 25 to 31 million barrels, will come from the Asia-Pacific region.[2] Does mushrooming energy demand from dynamic Asian economies portend a hastening of the end of the oil age, and with it, scenarios of conflict and competition for increasingly scarce hydrocarbon resources in the decades ahead? This is a matter of much debate in which the recent history of oil forecasting offers little reassurance. However, the reality on (and in) the ground suggests cause for at least cautious optimism.

The history of the petroleum era is littered with stunningly wrong predictions—almost since oil began to be commercially exploited a century and a half ago. In 1885, the U.S. Geological Survey said there was "little or no chance for oil in California." Six years later, it added Texas and Kansas to its list of states in which oil was unlikely to be found. In 1939 the Department of the Interior said only thirteen years of reserves were left; In 1951, twelve years later, the agency repeated the claim.[3] Yet for a generation after World War II, it was a time of plenty, of cheap oil and few existential concerns about energy security. Fears about the depletion of reserves, of that "Wolf at the Door," seemed to evaporate. Then, into the 1970s, booming demand caught up with supply and created a tight oil market, while postcolonial "resource nationalism" infused the Organization of Petroleum Exporting Countries (OPEC) with ostensible political purpose. The 1973 Arab–Israeli conflict offered the opportunity for both elements to combine into a potent force known as the Arab oil embargo. The result was the first oil shock of 1973–74, just a generation ago, when OPEC managed to flex its muscles and began to transform the world of oil with production cutbacks, price hikes, and oil embargoes.[4]

By the end of 1973, oil prices had risen from $1.80 a barrel in 1970 to $11.65 a barrel, a sixfold increase. In an exuberant, crowing mood, one of OPEC's most hawkish members, the Shah of Iran, warned that the industrialized nations would "have to realize that the end of their terrific progress and even more terrific income and wealth based on cheap oil is finished. They will have to find new sources of energy."[5] One of the top oil experts in the U.S. government, James Akins, concurred, going public with his concerns in an influential 1973 article in *Foreign Affairs* entitled, "The Oil Crisis: This Time the Wolf Is Here." So it seemed. By the end of the decade oil surged to nearly $50 a barrel as the 1979 downfall of the Shah, followed by the Iran-Iraq war the following year, sparked a second round of price shocks. Prominent industry analysts were predicting $100 a barrel oil, while emotionally charged speculation about the exhaustion of oil resources peppered the popular and industry press. (See Figure 2.1.)

It was in that tumultuous period of oil-driven economic instability, with

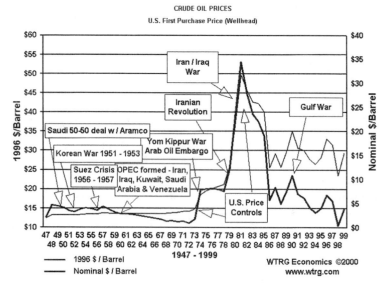

Figure 2.1 Crude Oil Prices
Source: WTRG Economics (www.wtrg.com), 2000.

The Myth of Energy Scarcity: Energy Crisis or Cornucopia? 19

angry and frustrated American consumers suddenly forced to line up at gas stations, that the specter of energy shortages, fear of dependence on imports, and a new depth of concern about energy security took hold. OPEC's success had the perverse consequence of encouraging investment in more marginal oil fields, which in turn, increased non-OPEC production, and no less important, accelerated the search for alternative energy sources—solar, geothermal, ethanol, nuclear, etc. Energy efficiency increased several times over, symbolized perhaps by the surge of Japanese subcompact and compact cars exported to the U.S. market and then the newfound ability of Ford, GM, and Chrysler (compelled by legislation and competition) to manufacture compact autos, many able to get 25 or more miles to the gallon.

The combination of enhanced efficiency and conservation measures along with radical changes in the functioning of what was becoming a truly global, and increasingly efficient, oil market led to reduced consumption. For example, between 1979 and 1983 alone, oil consumption in the noncommunist industrialized nations dropped by nearly 6 million barrels a day, from 51.6 to 45.7 million b/pd.[6] In the United States, as the economy moved from steel to software, a trend in many mature, postindustrial economies, the role of oil diminished sharply: between 1978 and 1998, U.S. GDP grew by 68 percent, but U.S. oil consumption in 1998 was 18.8 million barrels a day, basically the same as it was in 1978 (18.7 million).[7] This in turn led to a dissipation of the very concerns that helped generate this new set of circumstances.

Following the second oil shock in 1979–80, and throughout the 1980s, sagging prices and a new complacency were signs of the times. Even the 1990–91 Gulf War taking Kuwaiti and Iraqi production off the market only caused a small, and very short-term blip on the oil market radar screen. Indeed, a quarter century after the oil embargo, at the beginning of 1999, oil prices were hovering around $8 a barrel, in real dollars, significantly lower than they were in 1973. An unusually disciplined OPEC-led production cutback, unexpectedly rapid recovery of Asian demand, and a slowdown in investments in oil and gas exploration and production by major oil companies preoccupied with consolidating after a wave of megamergers pushed prices over $30 a barrel by mid-2000. In addition, a drawdown of oil stocks and limited spare capacity has led some analysts to conclude that a third oil shock and a price of $40 per barrel could occur, and that in any event, this short-term phenomenon would persist through 2001, masking longer-term trends.[8] Nonetheless, many analysts would concur with long-term forecasts by the IEA and the U.S. Energy Information Agency for prices to average in the range of $15–$20 a barrel over the next two decades.[9]

The "sky is falling" school of oil forecasting had, of course, been consistently wrong for more than a generation, much as its soul mates, Paul Ehrlich and Lester Brown, similarly continued to wrongly warn of population explosions and food shortages, respectively. Indeed, the Club of Rome in its dramatic 1972 *Limits to Growth* report said there were only 550 billion barrels of oil remaining, and that the world would run out by the end of the next decade. In fact, the world did consume 600 billion barrels of oil between 1970 and 1990. Yet proven reserves had risen to over 900 billion barrels by 1990.

That Wolf Again?

But a new factor was quietly beginning to gather momentum and international attention. It was a phenomenon part and parcel of East Asia's meteoric economic growth that become increasingly evident from the mid-1970s: an enormous surge in the region's need for energy, particularly oil. Indeed, for most of the past two decades, the locomotive of global oil markets has been burgeoning Asian demand, which grew nearly six times faster than that of Western industrialized states prior to the 1997 Asian financial crisis. Asian oil consumption more than doubled from 1985 to 1995, and is projected to double again by 2015.[10] Moreover, if China and India reached the same per capita energy use as the United States, global oil demand would increase roughly threefold.

More than any other single factor, this burgeoning Asian thirst for oil, a feature of the region's remarkable economic dynamism, sparked renewed apprehension about energy security in the latter 1990s after the lengthy interlude beginning in the early 1980s. By then, the trauma of the twin oil shocks of 1973–74 and 1979–80 dissipated into an eerie complacency as energy policy discussion seemed to fall off the foreign policy agenda entirely. But by the mid-1990s, many Asian political-military specialists began to voice fears that energy security concerns might turn the region's myriad territorial disputes into "resource wars."[11] Suddenly the issue of energy security, a constant front-burner concern in Asian nations such as Japan and Korea, began to resurface in the American public discourse. The April 1996 cover of the *Atlantic Monthly* bemoaning cutbacks in alternative energy research read "Drifting Towards Disaster." The article, one of whose authors was a former deputy secretary of energy, cited veteran energy hands from across the political spectrum warning that the next oil shock,

as one wag put it, "will make those of the 1970s seem trivial by comparison."
Over the next two years, a steady stream of forecasts of gloom and doom filled the intellectual landscape. Both the popular media and the elite journals were peppered with speculative new pessimism. A sampling ranges from the Sunday supplement *Parade* magazine asking "Could It [gas lines] Happen Again?" to *Scientific American* proclaiming "The End of Cheap Oil" and *Issues in Science and Technology,* a prominent science policy journal featuring "Heading Off the Permanent Oil Crisis" as prime examples of the persisting concern about energy security.[12] Or take the *Forbes* headline, "Cheap Oil: Enjoy It While It Lasts."[13] Then there was the *London Observer,* one of Britain's prominent Sunday papers, whose headline warned of an "energy apocalypse," proclaiming, "The world faces a devastating oil crisis in the early years of the new millennium . . . warnings of environmentalists in the Seventies appear now to have been prescient."[14] Former CIA director R. James Woolsey, writing with Senator Richard Lugar in *Foreign Affairs,* observed that "optimists about world oil reserves, such as the Department of Energy, are getting lonely." Their article details breakthroughs in cellulosic biomass (plants or plant products) that hold promise for making ethanol competitive as a clean, alternate transport fuel, a silver lining in the pessimist cloud (see below).[15]

Set against the aforementioned history of utterly wrong predictions about the end of oil, such admonitions, if made difficult to ignore by the gravitas of seasoned oil geologists and analysts, should be taken with a serious grain of a salt. The surfeit of gloomy forecasts by prominent oil specialists do not reflect any newly discovered facts or startling research finds. There are a plethora of complex factors—economic, technological, political, geological, and increasingly, environmental, that all impact current and future oil production. Most of the scarcity predictions are based on the same set of assumptions that have proven so flawed in the past.

Today, there are just over 1 trillion barrels of proven recoverable reserves (defined as recoverable at current prices under current conditions).[16] Indeed, the International Energy Agency (IEA) says there are 2.3 trillion barrels in ultimate recoverable reserves, and if unconventional sources such as tar sands and shale (more about them below) are included, the number may be well over 4 trillion barrels.[17] And these could well prove to be rather conservative figures, as the prospect of new fields combined with increasingly sophisticated technology that is able to extract growing proportions of what oil there is in wells continues to ratchet up inventories. Some estimates suggest substantially larger conventional oil resources in the range of 6 billion barrels, and unconventional oil resources at 14.9 trillion barrels.[18] The case against oil scarcity is strengthened by a recent landmark five-year long study by the U.S. Geological Survey, *USGS World Petroleum, Assessment 2000,* released in June 2000. The study concludes that there is an additional 20 percent increase in undiscovered oil, with the total of "future recoverable oil" *outside* the United States to be about 2,120 billion barrels.[19] USGS study said that in the Middle East, offshore areas of West Africa, and Latin America, there is significantly more oil than previously reported. In addition, the USGS study said that "reserve growth estimates nearly equal those of undiscovered resources" (which the study said totals 649 million barrels).

Gloomy Forecasters

Yet we are told, this time, the wolf really *is* coming. But not only are the scarcity forecasters—in some cases the same people who forecast hundred-dollar-a barrel oil by 2000 in the 1970s—still profoundly wrong. *But they have it exactly backward.* Indeed, such Cassandra-like noise obscures the far more likely prospect of an oil glut (not withstanding a volatile market and possible short-term price spikes and disruptions) and long-term relatively low oil prices in the first quarter of the new century. Being wrong over the past generation does not necessarily mean that scarcity forecasters are wrong now. While the exhaustion of fossil fuel resources is highly unlikely—certainly in the first half of the twenty-first century—at some point in the future, perhaps as early as the end of the twenty-first century, oil resources could well approach exhaustion.[20] But a close inspection of the forecasters' underlying assumptions reveals a chronic underappreciation of the role of technological change, and also its interaction with government policies (e.g., deregulation, privatization) and the Information Age-driven structural change in the operation of oil markets. Indeed, if oil production were judged to be on the downward slope, long before conventional oil resources dissipate, the accompanying rise in prices would almost certainly spur new investment in the development of alternative sources (see below) of fuel.

Conceding that dire projections of the 1970s were ill founded, two of the most prominent oil industry geologists, Dr. Colin J. Campbell and Jean H. Laherrere, argued in a *Scientific American* article that "the next oil crunch will not be so temporary. Our analysis of the discovery and production around the world suggests that within the next decade, the supply of conventional oil will be unable to keep up with demand."[21] These pessimists are not alone, and others, including at least one oil multinational CEO, ENI's Franco Bernabe, and several former senior U.S. energy and intelligence officials, tend to make roughly the same arguments. First, they challenge prevalent assessments of current reserves. These reserve figures, they argue, as compiled by leading sources such as the BP Amoco Statistical Review of World Energy, have been exaggerated by both companies and countries, the former, because it would add to their stock price, and the latter was inflated in the 1980s in order to grab a larger share of OPEC quotas. Colin Campbell argues, "There is thus good reason to suspect that when, during the late 1980s, 6 of 11 OPEC nations increased their reserve figures by colossal amounts, ranging from 42 to 197 per cent, they did so only to boost their export quotas."[22] The IEA reports that OPEC reserves increased by 62 percent from 1985 to 1989. ENI's Bernabe told *Forbes*, "It was simply a trick." Yet Campbell concedes that previous OPEC estimates were conservative, "so some upward revision was warranted." Campbell also takes issue with the way some reserves are calculated, particularly in countries that use less firm probability estimates than that required by the U.S. Security and Exchange Commission (90 percent, or P[probability] 90) Thus, this analysis leads to the conclusion that there are only about 1.8 trillion barrels of total recoverable reserves, roughly 25 percent less than the 2.3 trillion the International Energy Agency and others have estimated, almost half of which has already been extracted. (It should be mentioned, however, that the U.S. Geological Survey found that between 1980 and 1993 there were 925 reported discoveries of oil and gas that have not been added to reserve totals. This suggests that there is also some understating of reserves.)[23]

The second component of this argument is that there are no new major oil fields likely to be discovered, a conclusion that draws on trends in the rate of discovery of new fields, which has sharply diminished from 42 billion barrels a year in 1962 to 6.7 billion in 1998.[24] Moreover, 90 percent of current oil production comes from fields more than twenty years old, and 70 percent from fields more than thirty years old, according to Campbell.[25] This argument also assumes that further gains from technological innovation will be rather limited. These pessimists thus conclude that world oil production is projected to peak around 2003—meaning that more than half the world's oil will have been consumed—and begin to decline, sparking steadily rising prices, leaving the world first in the clutches of Gulf oil producers, and then out of oil altogether well before the end of the twenty-first century. The logic of this school of thought is summed up neatly by Campbell:

The world is using up its geological endowment of oil at a prodigious rate, and that rate will increase as newly wealthy countries, particularly in Asia, enter the industrial phase of economic growth. . . . At the same time, and despite astounding advances in the science of geology and in techniques of finding fossil fuel deposits, discovery rates of new oil reserves are falling sharply.[26]

The Campbell argument reflects two classics of the business of forecasting about natural resources. First is the Hotelling argument about resource scarcity, based on Harold Hotelling's 1931 paper, "The Economics of Exhaustible Resources," which put forward a "diminishing returns" case: that because the amount of resources are fixed stocks, as they are exploited there is less remaining, and thus the price tends to rise. But the crux of the Campbell argument is based on the Hubbert bell curve model, which predicted the decline in U.S. production in the 1980s. The theory of M. King Hubbert, a Shell Oil geologist in the 1950s, more generally was that because the amount of oil is finite, the future price of oil is an inclining curve, rising as the flow of oil slows and oil consumed becomes more dear than that remaining in the ground.[27] Thus, Campbell argues, the world will soon come to and pass its peak production rate (he suggests by 2003), bringing about a sustained rise in real oil prices over time.

Battle of the Forecasters

But the imminent scarcity argument has been offered before, not only in the 1970s but also again in the 1980s and early 1990s. In fact, Dr. Campbell himself has been predicting it for as many years.[28] In a kind of battle of the forecasters, Michael C. Lynch and M. A. Adelman, two prominent oil economists of the Massachusetts Institute of Technology (MIT), have become a kind of counterpoint to the petro-pessimists, particularly the views of Campbell. They essentially have a dynamic rather than a static view of hydrocarbon resources. Rather than viewing the amount of oil as a fixed number like the balance in a bank account—each dollar withdrawn means there is one less to draw on—they view it more as current inventory, a flow of reserves rather than a fixed stock. According to Adelman, "Reserves are inventories constantly used up and replaced."[29] It is worth noting that the ratio of world proven reserves to production, currently at 42 years—even after falling modestly since 1990—is still substantially higher than it was in 1972.[30]

Reserves, argues Adelman, are not simply "discovered" but created by investment. Indeed, in 1999, the investment in the oil industry was $83 billion.[31] It is price that is the indicator of scarcity, and "to explain the price of oil, we must discard all assumptions of a fixed stock and an inevitable long-run rise . . . ," according to Adelman. Thus, if the cost of developing and extracting reserves is less than the price of oil, companies will continue to invest in new reserves—and the cost of developing new oil reserves has plummeted dramatically over the past quarter century. By this logic, when the prospect of scarcity, and thus steady price hikes, begins to appear on the horizon, investment will flow to alternatives (e.g., tar sands, shale, ethanol) or new energy sources. Adelman and other economists, such as Julian Simon, question whether the notion of considering energy resources as finite is even a useful concept for economic analysis.[32]

The steady downward price slope of most minerals, and also of oil—to the degree that replacement costs have not increased—would tend to reinforce such logic. Indeed, looking at the estimates of reserves over the past three decades, one finds not only that global reserves have outpaced production, but the total proven reserves have virtually doubled since 1970! Of course, there is reason to believe that both countries and companies may exaggerate reserves. Nonetheless, the trend still holds. On the other hand, reserves in oil fields have generally tended to grow over time. In the United States, cure oil discovery peaked in 1930, with proved reserves at 13 billion barrels. Yet over the next six decades, the United States (not counting Alaska) produced 130 billion barrels, and now has some 17 billion. Similarly, an expert mission estimated Persian Gulf reserves in 1944 at 16 billion barrels proved and 5 billion probable.[33] Today, Persian Gulf proven reserves are estimated at nearly 700 billion barrels.[34] However much political manipulation of reserve numbers may have occurred, it is also true that reserves have continued to grow substantially over the past three decades.

One of the most glaring errors of oil forecasters has been a large underestimation of the oil potential in areas outside of OPEC. Over the last two decades, pundits predicted that production from non-OPEC sources would decline by 3.6 million b/pd or about 13 percent, leading to a gradual real increase in oil prices. Instead, non-OPEC production rose by over 4 million b/pd or about 15 percent. From 1978 to 1996, twenty-two additional countries began to produce oil.[35] In 1998, non-OPEC production exceeded that of OPEC—35 million bp/d or 47.7 percent of world total versus 30.7 b/pd or 42.1 percent of world production.[36] Problems in the Russian and Chinese oil sectors and a sudden collapse in oil prices combined to reduce non-OPEC production by 800,000 b/pd in 1998, supplies began to rise in 1999, and they will likely rebound over the next decade or so as the oil industry continues to perfect its methods of exploration and extraction. Ample new moderate-cost conventional reserves are planned to be brought on line in the coming years in areas such as offshore U.S. deepwater, Gulf of Mexico, offshore Africa, onshore central Africa, South America, eastern Canada, and frontier areas in the FSU and arctic reaches. (See Figure 2.2.)

Against this backdrop, Lynch has methodically traced and compared the track record of price forecasts over the past two decades, particularly those of Campbell, as one of the most prominent pessimists.[37] For example, in 1989 Campbell wrote an article in *Noroil,* a oil trade publication, arguing that while most forecasts see stable prices with a modest increase to the end of the century, "the slope of declining supply will become a precipice if production is held at present levels, never mind increasing. When that happens, a major leap in prices seems inevitable unless the major exporting governments exercise deliberate restraint. . . . Shortages seem to be inevitable by the late 1990s. . . . " Campbell's figures show world oil production peaking in the late 1980s and dropping from about 58 m b/pd to 45 m b/pd, predicting that the perception of declining reserves would lead to $50 a barrel oil by 1994.[38]

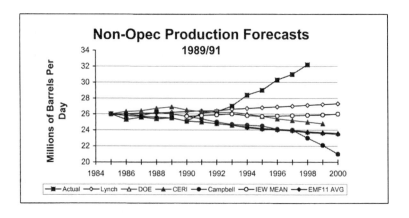

Figure 2.2 Non-OPEC Production Forecasts
Source: Michael C. Lynch (MIT), "Crying Wolf: Warnings About Oil Supply," March 1998.

Given the pessimist argument that there are no giant fields remaining to be discovered, one might expect predictions in mature producing areas with good databases such as Canada and the United States to be a fair measure of his assumptions and methodology. Yet charts prepared by Lynch illustrating Campbell's 1991 forecasts show extremely large errors, consistently more errant than other leading forecasters (see Table 2.1).Where Campbell has Canada declining from about 1.5 m b/pd to just over 500,000 by 2005, in reality Canadian production has shown a steady increase to over 2.6 m b/pd by the end of 1999. Similarly, for the United States, Campbell shows production on a downward slope from about 9 m b/pd in 1991 to less than 3 m b/pd by 2005. In fact, U.S. production had tapered off only slightly to just under 8 m b/pd by 1999.[39] As Lynch argues, ". . . both countries have outperformed his forecast far beyond levels conceivable to him. If this methodology doesn't work in mature regions with high-quality data, it is hardly likely to be reliable elsewhere."[40]

Examining remarkably similar oil forecasts made by Campbell in 1989, 1991, and 1997/98, all of which were widely off the mark, Lynch, whose own 1989 predictions of downward prices were dubbed "heretical" by a leading trade newsletter, persuasively argues that the pessimistic bias results from flawed methodology of the Hubbert curve. "This method," explains Lynch, "almost always produces a near-term peak [in the curve], no matter

Year	Base 1997	PIRA 2005	PIRA 2010	PFC 2005	PFC 2010	USDOE 2005	USDOE 2010	USDOE 2015	USDOE 2020	Purvin & Gertz 2005	Purvin & Gertz 2010	Purvin & Gertz 2015	IEA 2010	IEA 2020
Demand														
Western Hemisphere	27.4	31.7	34.0	32.3	34.3	32.9	36.1	37.5	39.9	30.9	32.9	34.6	32.4	35.1
Europe/FSU	20.1	22.2	23.5	21.7	22.1	21.3	22.7	24.2	25.5	22.4	24.2	26.3	24.2	27.2
Asia	19.6	23.9	27.7	22.0	23.6	24.8	28.5	33.0	38.4	26.0	28.9	31.7	29.0	37.5
China	4.0	5.9	7.4	5.4	6.0	5.6	7.0	8.8	11.2	5.9	6.9	8.0	7.1	10.1
Africa/Middle East	6.7	8.2	9.5	7.5	8.5	8.7	9.7	10.9	12.2	7.6	8.3	8.9	8.2	10.3
Total	73.8	86.0	94.7	83.4	88.6	87.7	97.0	105.6	116.0	86.9	94.3	101.5	93.8	110.1
Production														
Western Hemisphere	18.9	25.9	29.6	22.4	22.5	24.9	26.2	26.6	26.6	24.5	26.1	27.2	19.0	17.5
Europe/FSU	14.6	15.6	16.6	16.7	18.5	18.2	20.0	19.8	19.9	16.4	18.4	19.1	14.7	12.2
Asia	7.7	8.2	8.4	7.9	7.5	8.8	9.3	9.2	8.9	8.8	9.2	9.5	6.4	4.7
China	3.2	3.5	3.7	3.2	3.3	3.5	3.6	3.6	3.5	4.1	4.3	4.5	3.2	2.0
Africa/Middle East	31.9	39.1	44.5	36.6	40.8	36.4	41.8	51.8	63.3	35.7	39.4	45.0	52.5	55.5
Total	73.1	88.8	99.1	83.7	89.3	88.3	97.3	107.4	118.7	85.4	93.1	100.8	92.6	89.9
OPEC Production	30.3	32.6	37.8	41.4	47.3	34.2	40.6	49.9	60.5	NA	NA	NA	NA	NA
Non-OPEC Production	42.8	53.5	57.4	42.2	42.0	52.0	54.9	55.2	55.4	NA	NA	NA	NA	NA
Regional Balances														
Western Hemisphere	-5.1	-5.8	-4.4	-9.9	-11.8	-5.0	-9.9	-10.9	-13.3	-6.3	-6.7	-7.4	-13.4	-17.6
Europe/FSU	-5.5	-6.6	-6.9	-5.0	-3.6	-3.1	-2.7	-4.4	-5.6	-6.0	-5.8	-7.1	-9.5	-15.0
Asia	-11.8	-15.7	-19.3	-14.1	-16.1	-16.0	-19.2	-23.8	-29.5	-15.5	-18.4	-21.2	-22.6	-32.8
China	-0.8	-2.4	-3.7	-2.2	-2.7	-2.1	-3.4	-5.2	-7.7	-1.8	-2.6	-3.5	-3.9	-8.1
Africa/Middle East	26.0	30.9	35.0	29.1	32.3	27.7	32.1	40.9	51.1	28.1	31.1	36.0	44.3	45.2

Note: Some forecasts handle non-OPEC liquids and NGLs differently.

Table 2.1 Regional Oil Forecasts to 2020
Source: Amy Jaffe and Ronald Soligo, "China's Growing Energy Dependence: The Costs and Policy Implications of Supply Alternatives," James A. Baker III Institute for Public Policy, Rice University, 1999.

when or where it is applied, and thus constantly needs to be revised upwards." Campbell's predictions that the price of oil is about to increase because major non–Middle East oil producers are reaching the depletion point in the curve and global production cannot be significantly increased are characteristic of all his forecasts. Lynch concludes that all Campbell's forecasts contain the same argument, ignoring past failures, and that he has tended to "update his data, increase resource estimates and production forecasts and move his production peaks higher and further out."[41]

The intriguing question is why these and similar forecasts have been so far off the mark on actual prices and actual production? The answer seems to be that Campbell starts from a static view of what constitutes recoverable resources. Thus, as Lynch argues, when the estimate of the area under the curve (resources) is increased, the entire increase must be applied to future production. When I asked a senior economist at one of the world's largest oil companies why he thought the pessimists' arguments were so consistently wrong, he replied, "What they are always missing more than anything else is technological change."

The Oil Revolution

The oil and gas industry does not tend to evoke the public image of being part of the high-technology sector of the economy, or even part of the knowledge-based "new economy." The image is one of grimy workers drilling holes in oil fields, not of geologists and engineers sitting in front of a computerized command center controlling drills and production equipment in distant oil fields via encrypted satellite communications. Yet in fact, over the past quarter century, the exploration, extraction, and marketing of oil and to some extent, gas have undergone remarkable transformations, indeed a technological revolution, putting the industry on the cutting edge of the Information Age global marketplace as well as of computer-driven advances in geological science. These developments help explain the lack of any substantial rise in oil prices since the early 1980s and why many forecasts for future oil prices project them to remain within current ranges ($15-$25 barrel) over the coming two decades.[42]

Continuing trends toward deregulation and privatization have facilitated profound changes in the workings of oil and gas markets that have a direct bearing not only on price and supply but also on the new geopolitics of energy. Since 1980, when crude oil prices were decontrolled in the huge U.S. market, international crude oil markets have become increasingly global and transparent. This trend has been facilitated by the advent of oil futures, forward and derivative markets. Oil is no longer sold mainly through exclusive, secret long-term fixed price contract arrangements among a handful of major suppliers. Rather, oil price has been forced into the open, and oil and oil products are mainly sold on a free-market, floating price basis among a multitude of players, including financial institutions that have no stake in oil-industry assets.

In this commodified market, arbitrage generally generates similar, highly transparent prices worldwide, correcting for quality of oil product and differential transport costs. In this new, more efficient oil market, oil export movements of crude oil cargoes are driven mainly by transportation economics, with most sellers striving to sell their oil to the closest end-user market to maximize revenues. Smart companies can protect investment in these areas by selling production forward in new sophisticated financial markets for oil, using long-range derivatives and other instruments to lock in profits no matter what short-term policy OPEC implements. An energetic trader in Kuala Lumpur, Manila, or San Francisco could locate a cargo of Brent crude oil on his or her computer screen at below market price and save large amounts of money for their firm.

This new commodified market, where transport costs tend to be more a determining factor than political relationships, has led to significant changes taking place in the destination of oil flows from the Middle East. As we enter the twenty-first century, the clear trends are toward an increasingly bifurcated global oil market. Oil production increases in Venezuela, Colombia, Canada, Mexico and the U.S. deepwater Gulf have begun to crowd Persian Gulf oil out of the U.S. market. When growing oil imports from the Atlantic Basin (North Sea, West Coast of Africa) are added in to those from the Western Hemisphere, the divide appears sharp. The United States is increasingly able to rely on oil from the Western Hemisphere and Atlantic Basin. By 1999, the Persian Gulf accounted for less than 10 percent of total U.S. oil consumption. At the same time, East Asia is becoming steadily and inexorably more reliant on imports from the Persian Gulf, particularly as East Asian oil production approaches its peak and modest, sustained growth demand rebounds over the next two to four years. These new patterns of oil supply flows are changing both the nature and geopolitics of the oil market. Yet in a globalized oil market, where a disruption anywhere means a price hike everywhere, and where oil is fungible, the notion of "energy security" produced by the oil crises of the 1970s may require rethinking.

However, the new patterns of oil flow do not diminish the strategic importance of the Persian Gulf to the world oil industry. Indeed, the Gulf, with some two-thirds of global reserves, will almost certainly loom still larger in the oil equation in the next quarter century. Investment trends in the Middle East also bode well for a lower oil price scenario. Iraq, with its 110 billion in proven reserves, has been held back from oil markets by United Nations sanctions for almost a decade. One can expect to see a doubling of that country's production rates within five to ten years should a change in government or other factor lead to an easing of economic sanctions. Libya appears to be a similar case. And Iraq is unlikely to be the only one expanding capacity in the oil-rich region. Iran has begun to reopen its oil sector to foreign investors with the goal to raise output capacity—and with it political clout and national security—substantially. Saudi Arabia has also begun to open discussion with U.S. multinational oil firms about direct investment in its oil industry, seeking to protect its eminent position, stimulate its economy, and deepen economic ties to the United States. Kuwait is also in the exploratory stage of opening its oil sector to foreign investment.

Should planned Persian Gulf expansions take place in Kuwait, Iran, and Iraq and non-OPEC production grow at only two-thirds the rate it did between 1985 and 1995, oil markets could be even more oversupplied than in 1998, when prices collapsed to $8 a barrel. To consume all the oil that might be planned for investment over the next ten years, world oil demand would need to grow by more than 3 percent per annum over the period instead of the 1.8 percent growth per annum rate seen between 1980 and 1995.[43] Certainly, one can envision "surprises" in the form of short-term disruptions resulting from political turmoil in the Gulf (e.g., instability in Saudi Arabia or Iran). But there appears to be a compelling case for adequate supplies of petroleum at prices under $30 a barrel into the first quarter of the new century—with a high probability of $20 a barrel or less for most of that period.

Technological Revolution

Moreover, the surprising reality of relatively minimal declines in production from mature fields in the United States and North Sea, as well as the prolific development of oil resources close to the United States and far from the Persian Gulf, can be linked in part to another aspect of the transformation of the oil and gas industry: remarkable technological breakthroughs in the international hydrocarbons drilling and production business. Improvements in underground seismic imaging systems have had dramatic effect, driving large discoveries in the Western Hemisphere,

Africa, and North Sea. What analysts missed in expecting the non-OPEC phenomenon to be short-lived was that technological advances would significantly lower the costs of developing marginal reserves and improve the chances for new discoveries while at the same time enhancing recovery rates from the existing resource base.

One important advance has been the development and now standard use of three-dimensional (3–D) seismic analysis. This technique allowed seismologists to develop detailed three-dimensional maps of the oil trapped deep inside layers of porous rock, increasing the operator's knowledge of the oil reservoirs and hence their ability to locate and rate of recovery by 20 percent.[44] Seismic capability was a major force in the 1995 offshore U.S. discoveries in the Mississippi Canyon region, for example. New discoveries under salt layers previously blocked to oil and gas explorers could contribute between 100 and 300 million barrels a day of oil equivalent in the coming years. The use of 3–D seismic not only minimizes the incidence of dry holes but also reduces costs of missing a field that might otherwise not have been found. New four-component seafloor seismic systems allow explorers to see geological features below layers of gas, further improving geological data. The recent development of four-dimensional (4–D) seismic (the added dimension being time) tracks the movement of oil, gas, and water in the subterranean strata of drilled wells, allowing oil to be extracted still more efficiently and more cheaply.[45]

It is a remarkable marriage of advanced oil hardware and information technology that has made possible real-time virtual exploration. These new computer-aided exploration techniques and breakthroughs in what is called directional drilling have made some reserves that were technically possible more accessible and commercially viable. New equipment allows engineers to shift a well from vertical to horizontal drilling several miles underground. New, sophisticated sensors have resulted in qualitative improvement in the accuracy of drilling. These sensors can detect oil, water and gas, and then allow the operator to steer the drill into position to obtain the maximum yield.[46] This new technology is leading to "smart wells," according to Columbia University scientist Roger N. Anderson, that in the near future will use computers and water monitors to detect the dilution of the oil stream by water.

Still more remarkable are breakthroughs in deepwater production that are making possible the exploitation of fields that lie thousands of feet below the surface. Moreover, new technologies, originally developed by the U.S. Navy are facilitating exploration below layers of salt and igneous rock that have in the past hidden oil fields from view. Drilling for subsalt reservoirs is expected to yield further oil booms in the Gulf of Mexico within the 2000–2005 period. A consortium of oil companies has drilled an exploration well in the Gulf of Mexico 7,718 feet deep, using a former CIA ship, the *Glomar Explorer,* as a drill ship, and new, more capable ships are under construction that may be able to sink drill pipes more than 10,000 feet. Already, improved platform design and construction has lowered the costs of deepwater exploration dramatically from $15 to $12 a barrel in the late 1980s to $6 to $4 a barrel currently, allowing for the development of over 1 million b/pd of new production in the deep waters of the U.S. Gulf of Mexico alone.[47]

This technological progress has dramatically lowered the cost for finding and producing oil and natural gas and provided energy consumers with ample, inexpensive supplies at a time when earlier forecasts had predicted a shortage would occur. It has extended the life of existing wells, and allowed the oil industry to double the amount of oil recovered—from about 30 percent to 50–60 percent or more in many cases. By 1998, technology lowered average U.S. hydrocarbon finding costs to around $5 a barrel, down from $15 in the 1980s. Drillers are experiencing four times the success rate in natural gas exploration and six times the success rate in oil exploration. Finally, scientists continue to refine theories of geological development—how oil is actually formed—with some investigating the possibility of replenishing reservoirs of migrating oil.[48] While many industry analysts are skeptical about this idea, if proven, such theories would not only put to rest the application of the Hubbert curve to oil forever but also challenge the idea of imminent scarcity—if not that of the finite character of hydrocarbon resources.

Unconventional Oil

Beyond the myriad technological developments expanding the amount and duration of conventional petroleum resources, there are enormous quantities of unconventional oil resources that are technically already feasible, but either not yet commercially viable or limited in commercial application for cost or other reasons. The impediments tend to be less technical than questions of price. Just as OPEC's success led to investment in new non-OPEC sources, if oil prices rise to sustained levels of $30 a barrel or more, a whole range of new fuel sources are likely to emerge as commercially viable. Already, the vast oil sands of Alberta in Canada, with an estimated 1.7 trillion barrels of tarlike, heavy crude reserves have begun to be exploited. Suncor, a Canadian company, has developed sophisticated techniques by which the tar sands are mined and refined to oil products at $14 a barrel. More recently, a host of major U.S. multinational energy companies are beginning to invest in these fields as well.[49]

Another promising prospect is emerging technical advances in converting clean and plentiful natural gas into liquid fuels. The basic technology to achieve this has been in existence for more than half a century. Estimates are that there is enough currently recoverable natural gas in the world to produce between 500 billion and 1.6 trillion barrels of synthetic crude oil.[50] It is also possible that still larger amounts could be found in coal seams—coal bed methane is already beginning to be developed in several coal-rich countries, most recently by Texaco in China. Yet gas in many hydrocarbon-rich areas of the world has been so impractical that until very recently it has tended to either be burned off at the well site or reinjected back into the ground. Transporting gas long distances is economically uncompetitive, and it is far more expensive than oil to ship through pipelines. Cooling and compressing gas into liquid—liquefied natural gas—can be and is shipped by tanker, but the infrastructure needed to handle it requires expensive investments, and is complex and difficult to manage. But there are a whole range of new and emerging techniques that may make it possible to convert natural gas into a liquid form at room temperature and, convert a higher percentage of gas to liquid oil products, making it cheaper to produce and more economical to transport.[51] Such technical breakthroughs could make gas-to-liquids more economically attractive over the coming decade even if oil prices stay in the historically low ranges as discussed above. But if oil prices rise to the level of $25–$30 a barrel or above, and government and private sector planning becomes based on such price levels, then the prospects for widespread commercial application would loom very large.

Gas-to-liquids, however, is only one technology that is likely to change the global energy landscape in the twenty-first century. Ethanol, already produced (with heavy government subsidies) in the United States from corn and in Brazil from cane sugar (3 billion gallons a year), has already demonstrated its utility as a transportation fuel, if not yet its economic viability. Another promising technological development, this time in biotechnology, opens the possibility of turning the production of ethanol from biomass into a potential alternative to petroleum and/or to make it competitive as a complement to petroleum (e.g., blended with gasoline). Newly developed processes using biocatalysts—genetically engineered enzymes, bacteria, and yeasts—can significantly lower the costs and make it possible to use any plant or plant product (cellulosic biomass) to produce ethanol.

New genetically-engineered enzymes being developed are likely to yield commercial breakthroughs as early as this decade.

This technology could have tremendous benefits, particularly for developing countries like India or China, where farmers would have a new cash crop from agricultural wastes and countries a new locally produced source of energy. Apart from lowering import bills, if used, for example, to power stationary fuel cells, it could provide a new, clean, and decentralized local source of energy.[52] It would also have important environmental value as biomass ethanol produces 1 percent of the carbon dioxide of gasoline-powered autos or electric battery-powered cars.[53] Ethanol from biomass also has the advantage or requiring only modest adjustments to the existing transportation infrastructure. This new ethanol technology is being incorporated into a biomass plant in Louisiana by BC International, a firm that has patented the processes, and if the trend toward lower production costs (and which new enzymes could make possible) continues as industrializing this technology suggests, it could emerge as an important twenty-first century fuel source. Exploiting a large volume of gas hydrates, most in underwater seabeds, is another potential alternative fuel source under research.

Emerging Technologies—Green Cars?

Beyond alternate fuel sources, another development that will alter the energy picture in the first quarter of the twenty-first century is a quiet revolution under way in the automobile and energy industries. Driven by environmental and competitive pressures all the major auto manufacturers have developed next-generation, cleaner, more fuel efficient autos that use substantially less gasoline or no gasoline at all. At the Detroit auto show in 1998, the CEO of General Motors boldly predicted that "no car company will be able to thrive in the twenty-first century if it relies solely on internal combustion engines."[54] At the same time, Peter Bijur, Chairman of Texaco, made similar remarks. In an all but unnoticed speech, he argued that in the future, "we will see multiple ways to power cars—hybrids, advanced batteries, fuel cells, even cars that run on pure hydrogen." Such statements reflect the reality that major mega-firms like Texaco and BP Amoco are quietly investing in alternatives like fuel cells in an effort to hedge against the future. However skeptical they may be about the Kyoto Accord and global warming, the corporate giants' behavior suggests that new environmental concerns are acquiring a commercial logic.[55]

Environmental legislation requiring, for example, 2 percent zero emission autos by 2000 and 10 percent by 2003 in California and New York envisioned electric cars as the future. But electric cars, requiring a large, rechargeable battery, have a short range (100 miles), recharging problems, and limited consumer appeal and practicality—barring major technological breakthroughs. It has also led many U.S. states, like California, to alter their laws to adapt to lower emission vehicles (LEVs) that are beginning to show up on the road.

Instead, auto companies have made substantial investments in developing a variety of new auto technologies, with battery-powered autos only one (and less promising) in the mix, trying to position themselves to compete in whichever of the still uncertain technology that proves most commercially viable. Already, a new type of conventional gasoline motor, a direct-injection gasoline engine designed by Mitsubishi Motors that is both less polluting and uses 20 percent less fuel has been widely used in autos sold in Europe and Japan. This engine is capable of processing the type of low-sulphur diesel fuel being developed from gas-to-liquids technology discussed above.[56]

But virtually all the major global auto manufacturers—Toyota, Honda, Ford, GM, Chrysler—have developed prototypes of hybrid vehicles (HEV) and fuel cell electric vehicles (FCEV) as well as electric vehicles (EV), many of which will begin to be mass-marketed over the coming decade. Toyota has been selling some 2,000 hybrid models a month since 1998 of its hybrid (gas and electric) Prius in Japan. Ford has invested some $500 million in a joint venture with Daimler-Benz that hopes to market 100,000 sedan-size cars annually by 2004. Ford has unveiled a midsize fuel cell sedan prototype that weighs 1,300 pounds less than a similar-size Ford Taurus.

At present, hybrid technologies appear best positioned to begin the transition from internal combustion engines. They have the advantage of requiring the least change in consumer patterns, use gasoline or diesel, and drive similarly to current model autos. Unlike electric cars, hybrids need no recharging. The Prius model uses the gas motor for acceleration, while the engine automatically shuts off while idling, and when braking, the kinetic energy from the wheels turns into electricity to charge the battery. Advertisements for Toyota's Prius and Honda's new competing model, Insight, began showing up on the pages of *Time* and *Newsweek* in 1999. But the economics remain a question mark. In Japan, for example, the Prius initially sold for about $17,000, with Toyota losing as much as half its costs on each car.[57] Some analysts estimate each car has been subsidized by several thousand dollars. But in late 1999, Toyota announced plans to cut production costs and improve efficiency. Honda put its Insight on the U.S. market for under $20,000.

Improvements in production and economies of scale may well make hybrids economically competitive by the 2005–2015 period. Prius gets about 70 miles to the gallon, while Chrysler also has a 70 mile-per-gallon diesel/electric hybrid version of the Intrepid model and GM plans an 80 mile per-gallon model. The fuel cells use hydrogen (or liquid fuel) to combine with oxygen and release an electron that produces an electrical charge. The only emission on a fuel cell car would be a few drops of water from a plastic tailpipe. But fuel cells remain to be proven commercially adaptable to autos.

Many energy specialists, however, see fuel cells as likely to emerge as an important twenty-first century energy source, not only for transport but for stationary use to power homes, offices, and computers. Fuel cell technologies are especially promising technologies for developing nations that do not yet have as elaborate power infrastructures as many industrialized countries. Ballard Power Systems, for example, has developed a stationary fuel cell unit that by 2001 could be available for $3,000-$5,000. Another company, Plug Power, has also begun to install fuel cells to power commercial residences, and projects similar costs in the same time frame. As fuel cells have not yet begun to be fully mass-produced, costs are likely to drop dramatically over the coming two decades.

One could identify a plethora of other emerging technologies such as superconducting power lines, plastics-from-plants biotechnology, and fusion and wind turbines, all of which could well come to fruition over the coming half century. Large-capacity superconducting lines, for example, are planned by Detroit Edison to begin operating in mid-2000. If this technology proves economically viable, according to a Baker Institute report, it could revolutionize the transport of electricity from natural gas, replacing expensive pipelines with high-voltage power lines from power generation stations built near the gas fields, though such a development is probably at least a quarter century away.[58]

Similarly, hydrogen fusion could one day replace fission as the preferred form of nuclear energy. To date, only incremental progress has occurred, but it is conceivable that breakthroughs could take place providing an endless source of energy. One new design theory—a colliding beam fusion reactor—has been advanced by University of California and University of Florida physicists.[59] It would require a smaller and less expensive facility than previous fusion experiments. Another example of an obscure but potentially important energy source is a turbine that can turn ocean or river water into electricity, invented by a Russian émigré engineer at Northeastern University, Dr. Alexander Gorlov. Gorlov, who helped design the Aswan Dam in Egypt, believes he has created a clean form of hydropower as an alternative to large hydroelectric dams, one that he says is economically competitive when oil prices are $20–$25 a barrel on a sustained basis. Allied Signal, a large conglomerate, has licensed the invention and plans to install power farms on the coasts on several countries.

Energy Futures

These emerging energy forms and energy technologies-in-progress are outlined for illustrative purposes. It is impossible to predict which technologies will prove to be viable and commercial beyond a niche energy role, as appears to be the case with solar energy and windmills. But these technologies are merely a sampling of innovation in the field of energy that is a continuing process, and into which billions of dollars are regularly being invested. It is virtually inconceivable that none of such myriad projects will yield important fruits. Thus it is equally inconceivable that the universe of energy will remain a largely static one, with the world on a trajectory toward resource exhaustion. Whether or not the Kyoto Agreement on Global Warming goes into effect or not, concerns about global warming and other environmental considerations will drive innovation in the energy sector in the decades ahead. This is particularly true in Asia, where many of the "first generation" air and water pollution problems that Western nations have to a large degree cleaned up are only now beginning to be addressed.

It is worth pondering, for example, the implications of emerging transportation technologies discussed above. Whether is it hybrid, electric, or fuel cell, one or more of these post-internal combustion engine technologies will almost certainly become everyday fare for consumers over the coming two decades. When pressed on background, a senior auto company executive said that by 2020 he envisioned barely half the autos on the road being internal combustion engine models. According to the IEA, more than 50 percent of the growth in global oil demand between 1995 and 2020 will come from the transportation sector. While it is difficult to forecast the implications, one can imagine how widespread use of autos that get 80 miles to the gallon would require dramatic downward revisions in all existing forecasts for oil consumption.

It is a relatively safe bet that the world economy will remain dependent on fossil fuels during the first half of the twenty-first century. As we have seen, the revolution in the oil industry is likely to restrain prices, which will most probably gradually rise to levels rendering unconventional oil sources and alternative fuel sources more commercially viable over time. We have already begun to see in the mega-mergers such as BP Amoco-Arco and Exxon Mobil major oil companies begin to position themselves to develop into more diversified energy companies. Discussions with officials from several major oil companies confirm that their strategic planning for the next decade is based on developing a more diverse mix of energy sources. The economies of scale of the new oil conglomerates are well suited to be able to manage the investments, the venture capital required to exploit the emerging technologies over the horizon. It is quite possible, if not probable, that toward the end of the twenty-first century oil will gradually come to occupy a declining percentage of the mix of global energy consumption.

It is worth noting one rather sanguine study by Dr. Michael May of Stanford University, a former director of Lawrence Livermore Laboratory. Assessing the long-term Asian energy and security situation, May, with a strong technical background and nearly four decades' experience in analyzing security and energy issues, projected trends out to 2050 based on relatively optimistic economic growth forecasts and concluded:

> Energy supplies for East Asian economic growth, as well as for other anticipated world energy needs will be available over the period considered—to the middle of the next century—at prices that will not seriously set that growth back, either in East Asia or in developed countries, provided that (1) international markets, including but not limited to markets for fuels and capital, continue to operate at least as freely as they have; and (2) either through prices or possibly by other means such as state allocations, the investments needed to bring new supplies to market are made.[60]

In the interim, relatively low oil prices should not argue for complacency in either the private sector or governments. At the same time, fears of scarcity will also prove to be a poor guide for energy choices. Rather, energy choices should be made in a comprehensive and integrated manner weighing environmental concerns, technological trends, and security considerations. Above all, decision makers should maintain a healthy respect for the forces of the market. As we have seen, market forces have solved many of the energy concerns of the 1970s. In the end, it would be unwise to allow scarcity to animate decision making. Rather, it is a question of what mix of energy choices—oil, gas, coal, nuclear—on balance are best suited to each nation's requirements, and what is the most efficacious and equitable means of meeting those energy needs. This suggests a very different perspective on energy, indeed, a new paradigm rather than the psychology of scarcity. Yet it is this traditional notion that has been the dominant way of thinking reflected in the energy policies and the role of the state in many Asian nations.

As Asia-Pacific nations begin to come to terms with the energy challenges of a new century, the choices they make to do so will have a major impact not only on their ability to fulfill their respective energy needs but also on geopolitics in the region as well as global markets. Whether they gravitate—as some have already begun to do—toward market-based solutions and realize the myriad commercial possibilities of foreign investment, regional integration and privatization, and deregulation or older *dirigiste* models may be the difference between increased conflict or increased cooperation in Asia.

Chapter 3

The Myth of Caspian Great Game and the "New Persian Gulf"

If old pessimistic pseudo-verities of resource scarcity resurfaced in the late-1990s, a wholly contemporary fallacy of misplaced optimism also emerged: the idea that the newly independent states of Central Asia and the Caucasus ringing the Caspian Sea promise a resource bonanza, a new "Persian Gulf," with competition over control of oil and gas reshaping geopolitics into a twenty-first century version of the old "Great Game" between Britain and Russia in the nineteenth century. Such hopes, however, have fostered perceptions not supported by knowable facts, which have led to dangerous exaggeration of the region's commercial and strategic significance. Indeed, the fate of the Caspian region may be an interesting test of whether the commercial logic of geo-economics or the strategic logic of geopolitics prevails.

The meteoric rise of the Caspian region to center stage of world politics (and its ascendancy on U.S. and Asian foreign policy agendas) underscores yet again that oil remains "The Prize," as Daniel Yergin dubbed it in his Pulitzer Prize-winning epic history. Before the demise of the Soviet Union in 1991, the term Caspian Basin was rarely invoked even by Soviet experts. The Soviet republics of the Caucasus and Central Asia were the obscure preserve of "Soviet nationalities" specialists. Suddenly, with the breakup of the Soviet Union, new frontiers of prospective oil and gas presented themselves; Islamic territories that had only been confederations of clans and tribes before being absorbed into the Russian empire in the nineteenth century were overnight independent nation-states.

A part of the world until recently unknown and profoundly esoteric to 99.9 percent of Americans, Japanese, and Filipinos alike suddenly adorns the cover of the *New York Times* magazine, and even *Parade,* the popular

Sunday supplement in hundreds of U.S. newspapers.[1] And that ultimate sign of arrival: it boasts its own James Bond movie, the 1999 thriller, *The World Is Not Enough,* filmed in Baku, Azerbaijan, with a plot revolving around the struggle over which pipeline route should prevail. The Caspian region has similarly moved up the ranks of importance on the U.S. foreign policy agenda, going, in the case of the United States, from a backwater on the Soviet affairs desk to a trendy, career-enhancing assignment that rates its own presidential special envoy and inter-agency task force. As evidenced by Madeline Albright's April 2000 tour, it rates hectoring visits by the American secretary of states as now part of the diplomatic landscape. Leaders from Kazakhstan and Azerbaijan are ushered into the White House like new potentates. Indeed, when President Clinton attended a November 18, 1999, signing ceremony in Istanbul of a preliminary agreement among states of the region creating a legal framework for a trans-Caspian oil pipeline the *New York Times* described it in a front-page story as "one of the President's most cherished foreign policy projects." The specter of the Chinese premier out-lobbying vice president Al Gore to get a Chinese state-oil company an oil concession in Kazakhstan underscores that the region holds a similar priority for Beijing.

The initial flurry of literature and official policy statements amplifies this view. "Central Asia," writes leading specialist S. Frederick Starr, "is once more a key to the security of all Eurasia."[2] Ian Bremmer, in an otherwise sober analysis claims, "the Soviet Union has been eclipsed by the Caspian Basin as an American strategic priority."[3] Still more remarkable is the doubly dubious claim of U.S. Secretary of Energy Bill Richardson, arguing, "The Caspian region will hopefully save us from total dependence on Middle East oil."[4] Yet similar views can be found in discussions with Chinese and Japanese analysts and in their respective newspapers and journals.

It is not difficult to see how the current obsession with the Caspian has gained its remarkable political momentum. In fact, the mix of wildcatters, major oil companies, and political jockeying bears far more resemblance to the initial scrambling for concessions in Saudi Arabia nearly 100 years ago, a classic tale of oil, money, and power, than the "Great Game" of the nineteenth century.[5] In this case, the sudden opening up of one the last unexploited, massive oil and gas fields sent major oil firms scrambling not to get left out. Local regimes, impressed by the alacrity of Western oil firms to curry favor, shrewdly realized that the Western companies' governments could provide assistance as a buffer against domination by Moscow. It seems to have been forgotten that the nineteenth century "Great Game" was not really about Central Asia but rather about British dominance of India. Central Asia was only of importance as a buffer for Britain against feared Russian incursions to displace the British in India.

The romantic dream that these newly independent states of the Caspian, wrested from the Russia yoke, could at once solve Western energy needs, expand democracy, and garner geopolitical advantages against Russia, Iran, and perhaps China may be alluring to underemployed or misguided post-Cold War strategists and oil executives alike. But it is a chimera that could prove counterproductive to American and even NATO long-run interests in Russia and Iran. It could also inadvertently draw the United States into a snake pit of local and ethnic conflict in unstable, embryonic states that Zbigniew Brzezinski has aptly dubbed the "Eurasian Balkans."[6]

Energy Bonanza?

Nonetheless, the allure of new, potentially huge oil and gas fields to add to their balance sheets understandably led the major Western oil companies to rush into the new energy-bearing states lest they miss an important opportunity. They quickly hired a star-studded cast of former Secretaries of State, National Security Advisors, and other ex-U.S. government officials to promote these new interests in the Caspian. This may have facilitated a similar fascination that gripped the U.S. government—important new interests to protect, new states to democratize, a whole new bureaucracy to create. One measure of this seemingly unbridled enthusiasm—bordering on self-aggrandizing hyperbole—was a report to Congress issued by the U.S. Department of State in 1997 putting the weight of the U.S. government behind a highly speculative—some might say fanciful—estimate that the region might hold "up to 200 billion barrels" of oil reserves. This figure was subsequently cited by top U.S. officials, including Deputy Secretary of State Strobe Talbot.[7] The U.S. government report sent the Western media into a feeding frenzy. The *Wall Street Journal* announced that Caspian Energy might be bigger than previously thought.[8] The *Economist* magazine expressed concerns about the rise of "Dutch oil disease" that was likely to accompany the flood of oil revenue.[9]

Suddenly, Kazakhstan was being touted as the next "Saudi Arabia." But amid all this excitement, little distinction was made between the potential reserves of the Caspian, resources that might be uncovered after perhaps two decades of exploration, and the immense proven reserves of the Persian Gulf—two-thirds of the world's total—that have already been discovered and delineated by long-standing, substantial drilling and seismic programs. Although it may be several more years before definitive judgments can fairly be made about the resource potential of the Caspian Basin, the region's resources are not of sufficient magnitude to dramatically alter the shape of global oil markets or displace the Persian Gulf.

The proven oil reserves of the Persian Gulf exceed 600 billion barrels. The proven oil reserves of Central Asia and the Caucasus are pegged at roughly 30 billion barrels, representing an oil province roughly the size, in terms of proven oil resources, of Norway or Libya, or less than 3 percent of the world proven oil reserves.[10] By comparison, proven oil reserves in the Middle East account for nearly two-thirds of the world's proven reserves. Moreover, where oil in Saudi Arabia can be produced for under $3 a barrel, the expensive transport costs of Caspian oil mean that oil companies need about $13 a barrel to turn a profit from Caspian oil exports.

But there is no question that the oil reserves of the Caspian Basin are significant even if not close to those in the Persian Gulf. Future exploration may confirm that the region potentially holds between 50 and 140 billion barrels of oil.[11] However, this figure remains speculative and even high-end estimates are not in the same league as Saudi Arabia with its 269 billion barrels of already discovered proven oil reserves. Several major Western energy companies now appear to be leaning toward lower-end estimates of total recoverable reserves. After an initial phase of exploration, the mood among a number of Western oil companies is decidedly circumspect. One somewhat disillusioned oil company official told the author, "In retrospect, the Russians didn't do such a bad job looking for oil, given the limits of their technology."[12] Much of the hope for new Caspian reserves centers on the Kashagan field in Kazakhstan, which seismic surveys indicate may be several times larger than that nation's huge Tengiz field, one of the world's largest. Western firms exploring the field suggested in June 2000 that Kashagan's potential appeared very promising. If Kashagan does hold 20–30 billion barrels of reserves (nearly doubling Caspian proven reserves) it could produce 1 to 1.5 million b/pd.[13] Such volume could moot the pipeline competition; an east-west (i.e., Baku-Ceyhan) as well as north-south pipeline would then become economically feasible.

But the size of the Caspian reserves—whatever they may prove to be, over the next 10 to 15 years—are only one part of the a much larger story and may not achieve the output potential its promising reserve estimates imply. Obscured by all the fanfare are a daunting set of complex technical, economic, logistical, geopolitical, and social obstacles that must be overcome before the Caspian's potential can be realized. Perhaps the most fundamental overlooked aspect of the Caspian energy issue is one simple and unique feature: Caspian hydrocarbon resources are landlocked and located at a great distance from the world's major energy-consuming regions.

Landlocked: Drilling Problems

In fact, the geography of Caspian reserves may be one of the most complex in the history of the oil industry: rarely have major oil exporters faced such a situation with no direct outlets to foreign markets. The region's producers cannot simply ship oil by tanker from domestic ports to international sea-lanes as is done from the Arab Gulf. Instead, the Central Asian and Caucasus states must rely on costly pipelines crossing several neighboring countries as the chief means of transport. Moreover, none of the many possible routes under consideration offer even a whiff of reprieve from the region's hornet's nest of long-standing ethnic conflict and bureaucratic entrapments.

For openers, there are vastly underappreciated drilling constraints. The Caspian Basin is far from major supply centers for exploratory equipment and faces a debilitating shortage of modern drilling platforms and other related supplies. This shortage is more severe than virtually anywhere else. Despite huge demand for equipment, there are only two assembly yards equipped for manufacturing or refurbishing offshore drilling rigs for the region: one at Astrakhan in Russia along the northern Caspian and one in Primorsk, near Baku.[14] In early 2000 there were only two operational semi-submersible platforms operating in the region. To bring additional semi-submersible rigs into the Caspian Sea will be a logistically difficult and expensive venture, forcing the oil companies operating in the region to pool resources and take turns drilling prospects. The usual practice of towing platforms out to sea is not possible. Oil rigs have had to be cut up into parts, floated down the Volga River, and then reassembled. BP Amoco is reported to have spent nearly $200 million to reconstitute one such rig. To put it in perspective, the North Sea, with comparable reserves, had nearly 100 offshore rigs in operation as it tripled production from 1980 to 6.1 million b/pd in 1998.

Such constraints severely limit the amount of drilling that can take place in the region at any one time. They also mean oil well installations take considerably longer to finish—in some cases up to two years as compared to two to three months in many other oil provinces in other parts of the world. These constraints also mean that a discouraging first effort can have a devastating impact on development schedules for any particular field. If an oil company has a dry hole or expected natural gas find, rather than just drill again at another location, it may have to get into the queue before a second rig can be acquired to drill elsewhere in its exploration concession area. In the case of Azerbaijan's Karabakh field, for example, initial failures will mean that schedules to produce oil at the field by 2001 are no longer feasible.

Drilling obstacles also mean that while its energy resources may be geologically equivalent in scale to the North Sea, the region's output is unlikely to achieve such potential. North Sea production rose from roughly 2 million b/pd in 1980 to 6.1 million b/pd today, or 8 percent of current world demand. By contrast, after two decades of development, Caspian oil production may top no more than half that and represent little more than 3 to 4 percent of world oil demand by 2010. Projections for Caspian production by 2010 range as high as 4.5 million b/pd. But at present, unless oil prices stabilize at high levels, the IEA "low case" scenario—as little as 2.5 to 2.8 million b/pd by 2010—appears a higher probability. Of those exports, perhaps as much as 1 to 1.5 million b/pd could remain within the Black Sea region. By 2000, however, the Caspian was producing just 1.2 million b/pd, less than 200,000 of which were exported outside the immediate region.

In fact, a series of dry holes drilled in late 1998 and 1999 in Azerbaijan, combined with a large BP Amoco gas discovery at its Shah Deniz field, led two major Western firms to forgo their oil concessions and withdraw their operations from Azerbaijan in 1999. Azerbaijani gas may require Baku to reach a commercial understanding with Turkmenistan, which possesses the world's fourth largest gas reserves and whose pipeline to markets in Turkey must pass through Azerbaijan. This situation has complicated an already problematic U.S.-led efforts to build a major east-west pipeline from Baku to Ceyhan on Turkey's Mediterranean coast. Baku-Ceyhan, as it is known, appears the least commercially competitive of several possible oil and gas pipeline schemes, one that the Western-dominated oil consortia who control the production of the vast majority of Caspian hydrocarbon resources have viewed with considerable skepticism.

Landlocked: Transport Dilemmas

The Azerbaijan-Turkmenistan pipeline sharing question and the debate over Baku-Ceyhan offers hints as to why the logistical problems involved in drilling for oil are eclipsed by the still more daunting transport problems involved in getting the oil and/or gas to market. Not only is the region landlocked, but the major Caspian resource-states, Kazakhstan, Azerbaijian, and Turkmenistan are surrounded by neighbors who are commercial rivals and possess the major access routes. For example, Russia, with the most elaborate, if dilapidated, pipeline network, has traditionally dominated transit routes for its former republics. Thus, in a dispute with Turkmenistan, Moscow cut off the new state's pipeline flow to European markets in 1997. Iran, by dint of geography, possesses the shortest, most commercially attractive access routes.

All told, a plethora of alternative oil and gas pipeline routes have been proposed, and current political gaming in the region revolves around the "division of rents" (e.g., tariffs and transit fees to be garnered) and geopolitical influence a firm decision on pipeline routing will generate.[15] Each of several east-west and north-south pipeline routes has its own political calculus and economic logic. The United States has waged an impressive and sustained diplomatic campaign, investing enormous political capital in order to make Baku-Ceyhan the Main Export Route (MER), the course favored by Azerbaijan, and especially by Turkey, and more broadly, to mold the region in its image. The key strategic point about Baku-Ceyhan is that is would skirt both Russia and Iran, while reinforcing Turkey's economy and role as an oil entrepôt.

Washington has multiple objectives that it has repeated mantra-like, with a few minor variations, in virtually every official policy pronouncement: strengthening the independence and promotion of democracy of the newly independent states; increasing and diversifying world energy supplies; enhancing commercial opportunities for U.S. companies; bolstering the energy security of the United States and its allies and friends.[16] Decoding this diplomatic vocabulary and the geopolitics of this policy means freeing Central Asian and Caucasus states from overdependence on Russia, continuing to isolate Iran, and weaving Central Asia and the Caucasus into the fabric of the international economic and political system. Secretary of Energy Bill Richardson articulated one of the most lucid, unvarnished descriptions of American policy in an interview with the *New York Times:*

> This is about America's energy security, which depends on diversifying our sources of oil and gas worldwide. It's also about preventing strategic inroads by those who don't share our values. We're trying to move these newly independent countries towards the West. We would like to see them reliant on Western commercial and political interests rather than going another way. We've made a substantial political investment in the Caspian, and it's very important to us that both the pipeline map and the politics come out right.[17]

Unfortunately, the very headline in the *Times* article in which Richardson is quoted hints at some of the inherent problems, if not contradictions in U.S. policy: "On Piping Out Caspian Oil, U.S. Insists the Cheaper, Shorter Way Isn't Better." There are several other oil and gas pipeline routes, both east-west and north-south. (See Figure 3.1.) The shortest, least expensive route is a pipeline south to Iran, with oil to be shipped out from transport lines in the Persian Gulf. There are variations on these pipeline options for gas and oil, including still shorter pipelines to feed demand in northern Iran with Tehran swapping its own oil for export. There is a northern route into Russia and its existing pipeline system as well as a northwest route to the Russian port of Novorossiysk for shipment to Europe through the Black Sea. This route has operated to move "old oil" Pre-Soviet break-up), when problems such as conflicts in Chechnya and Dagestan have not disrupted operations. Another east-west pipeline, albeit modest in scale, already operates taking about 100,000 barrels a day from Baku through Georgia (bypassing Russia) to the Black Sea port of Supsa. In addition there are two west-east pipeline ideas, both from Turkmenistan, with one through Afghanistan to Pakistan and India; and a second through Kazakhstan to China. Both ideas are dormant as of 2000.

Understandably, the leaders of the fledgling states of the region have powerful and compelling financial reasons for a quick decision on how to transport their precious energy resources to market. All of the Caspian Basin states have fragile economies and view oil and gas revenues as critical not only to their well-being but in some respects their viability as nation-states. But there are also compelling arguments against a choice made in haste. There are a number of long-range variables that will influence what will principally be commercially driven decisions (it is the oil consortia who must pay for the pipelines). These variables range from shifts in Russian attitudes and behavior toward its Republics and the West, an economic revival in former Warsaw Pact states and the Ukraine that could create bigger local markets for oil and gas, and the geopolitical and commercial implications of rapidly evolving political change in Iran.

But the most fundamental reason for cautious, prudent decisions is that the approximately one million barrels a day Azerbaijan International Operating Company (AIOC) says is necessary for the Baku-Ceyhan pipeline to be profitable has yet to be found! A major find at the Kashagan field in Kazakhstan may provide sufficient amounts of crude for the large pipeline to go forward. Indeed, oil industry sources suggest that the new finds at Kashagan may be three to four times the size of Tengiz, the largest oil field in the region. If this is the case, it might yield 1 to 1.5 million b/pd by 2001–2002. This would mean a total of some 30 billion barrels in recoverable reserves, roughly doubling existing estimates of Caspian reserves but

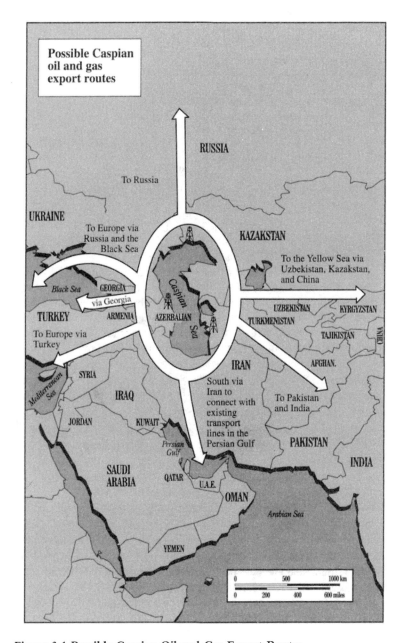

Figure 3.1 Possible Caspian Oil and Gas Export Routes
Source: Geoffrey Kemp, "Energy Superbowl: Strategic Politics and the Persian Gulf and Caspian Basin," (Washington, D.C.: The Nixon Center, 1997).

still several orders of magnitude smaller than the official State Department report has suggested.[18]

But it remains uncertain what the total volume of oil and gas reserves of the Caspian Basin may be, and an authoritative assessment is not likely to be attainable before 2001–2002 or later.[19] The November 1999 Baku-Ceyhan accords signed with great fanfare in Istanbul only create a legal framework for moving forward with the pipeline, and the promise of a Turkish guarantee to cover costs beyond the $2.4 billion Washington and Istanbul estimate the price of constructing the 1200 mile pipeline. Oil companies such as BP Amoco estimate the construction costs of Baku-Ceyhan at $3.8 billion; other oil companies in the main AOIC surveyed by the author claim still higher estimates.[20] If Kashagan can produce 1 or, by some estimates, 2 million or more b/pd, not only would that provide economic feasibility for Baku-Ceyhan, but it might end the "mediocre game" of pipeline competition by creating an economic basis for other pipelines, not least, through Iran.

The cautious move by the oil consortia underscores the prematurity of making decisions with long-term consequences in a fluid political and uncertain economic environment based on judgments of short-term realities. Public rhetoric by the Clinton administration about the value of an east-west route from Baku to the Turkish Mediterranean port of Ceyhan notwithstanding, the United States has been unwilling to subsidize the pipeline, leaving the financial burden principally in the private sector. Both adequate amounts of oil to export and stable prices at above $13 a barrel are prerequisites for AIOC to give a green light to Baku-Ceyhan. In the interim, as one U.S. official conceded, "all the routes are still up in the air."[21]

Political Risk

In addition, there is a layer of political risk in the region's numerous ongoing local conflicts and disputes that must be part of the Caspian energy development calculus. Take for example the fact that both the Baku-Ceyhan and Baku-Supsa pipelines traverse the Republic of Georgia, where Moscow can assert itself by supporting that country's already simmering separatist movements—as it has done in the past. Georgia's leadership currently suffers under the heavy pressures of blatant Russian interference in its internal affairs. Bloodshed has already been spilled over Abkhazia's aspirations for independence while tensions are high in other regions as well, including South Ossetia, Ajaria, and the Javakheti region.[22]

As if the problems in Georgia weren't sufficiently discouraging, this preferred pipeline route and the more modest operational one also passes within artillery range of Russian-backed Armenian militias who are pitted in a deeply entrenched territorial conflict with Azerbaijan. Certain Armenian hard-line factions believe that it is vital that Azerbaijan be denied significant oil revenues that could be channeled into arms acquisition programs. This sentiment is entrenched despite the fact that Armenia currently maintains a sizable military edge over Baku. Dividing up transit fees and other benefits from oil and gas development may, as was the case in the previous 1996 conflict in Chechnya, eventually become a positive factor in resolution of the conflict.

But chances of speedy resolution of the Armenian/Azeri conflict appear remote. Azerbaijan's government faces a large-scale refugee problem deriving from Armenian occupation of Azeri territory during fighting between the two countries. These refugees tie the hands of Azerbaijan's president Haydar Aliev from showing flexibility. Meanwhile, the Armenian population of the disputed territory—Nagorno-Karabakh—cite historical claims to demand self-rule and autonomy or even statehood in association with Armenia. They are supported financially and politically by the Armenian diaspora in the United States and Europe and by the government of Armenia. Their U.S. lobby has been so successful that the U.S. Congress initiated sanctions against Azerbaijan in 1992. But there is growing support for lifting them if any progress can be made in conflict resolution between the two sides. The conflict obviously pours cold water on hopes of an export route from Baku via Armenia.

Other political barriers have also inhibited a steady stream of Caspian energy exports more directly through Russia or Iran. Moscow's bureaucratic chaos, corruption, and weak administrative control have combined to deny access to key sections of Russia's vast pipeline network. Moreover, certain elements inside Russia's foreign policy hierarchy work to block exports from Kazakhstan, Azerbaijan, and Turkmenistan on nationalistic and economic grounds.

Russia has attempted to retain control over Caspian oil developments by utilizing its leverage over transportation routes, asserting legal claims in the Caspian Sea and pressing Western companies to include Russia oil firms into international consortia. The view from inside Russia's oil sector—which through Lukoil holds a 10 percent stake in AIOC and 5 percent in Kazakhstan's Tengiz field—continues to be integrationist.[23] The former Soviet state oil sector invested heavily in existing Caspian oil facilities and Moscow's oil men believe this entitles Russia to control and compensation. Difficulties in obtaining land rights, transit rights, etc., may eventually lead to Russian firms being given larger shares as an incentive to pressure both Moscow and local authorities to cooperate.

In any case, there are competing Russian interests and tendencies, with a divide between nationalists who consider Central Asia their birthright and a constituency of corporate interests and more pragmatic economic realists gravitating toward a more cooperative posture.[24] However, Russia's economic weakness suggests it is more likely either to play a spoiler role in regard to the development of Caspian resources or seek to use the threat of being a spoiler to gain leverage to be better dealt in to what is a not so great game by the Western oil consortia.[25] More recently, there seems to be a countertrend of the Putin Administration reasserting itself in the region. Torn by instability and Islamic insurgency, many Central Asian states appear to be rethinking relations with Russia as the United States continues to focus more on democracy, as beleaguered regimes are looking for support for tough measures against prospective instability. The prospect of less conditional military support from a sympathetic Russia has a trend toward improving military and economic ties between Central Asian states and Russia. This was manifested in the welcoming of Putin to the region with open arms during his May 2000 visit.[26]

In contrast, despite its geographic advantages, Iran has had much less leverage to exploit in the pipeline game. Iranian export routes have largely been on hold as a result of the containment approach of the U.S. administration and the U.S. Congress that still views Tehran as a "rogue" regime to be isolated. Since the landslide election in 1997 of Mohamed Khatemi, a gradual diplomatic dance between the United States and Iran has slowly unfolded. This began with Khatemi's CNN interview in January 1998, and saw Clinton waiving sanctions against Russian, French, and Malaysian firms investing in Iran's oil industry in May 1999, under the extraterritorial Iran-Libya Sanctions Act (ILSA). That June, there was a diplomatic overture from Secretary of State Madeline Albright. In a speech to the Asia Society, Albright described Khatemi's electoral victory as "a mandate for change, demanding from the Iranian government greater freedoms, a more civil society based on rule of law, and a more moderate foreign policy aimed at ending Iran's estrangement from the international community."[27] But a backlash by conservatives who hold the levers of power makes the pace of change problematic.

Britain and other EU nations have begun to restore diplomatic relations with Iran, and Khatemi's apparent pulling back from the fatwa, the religious decree ordering the murder of Salman Rushdie at the U.N. General Assembly in September 1998, has dramatically softened the atmosphere.

Absent Iran's revolutionary evangelical ideology, there is no necessary reason for adversarial U.S.-Iranian relations. Moreover, while there is major internal political drama playing out in Iran, the overwhelming reformist victory in the February 2000 parliamentary elections underscore that Khatemi's ascendancy marks the beginning of the end of first-generation Iranian revolutionary leadership, if not the total waning of the revolution. Neither private Iranian specialists nor U.S. officials dispute the possibility that within five years or less Iran might be viewed as a neutral—if not friendly—nation by the United States.[28] Such an eventuality would recast the pipeline equation dramatically, with Western oil companies strongly favoring the shorter, less expensive routes through Iran for Caspian oil.

Almost all of the biggest oil producing consortia in the Caspian region contain at least one U.S. oil company partner and are thereby barred by U.S. laws from undertaking major energy investments in Iran. Japan has equity in AIOC as well. The U.S. government has also worked behind the scenes with mixed success to thwart foreign companies from joining with Iran's oil company NIOC to construct energy export outlets via Iran. Were sanctions to be lifted in a rapprochement between Tehran and Washington, the attractions of various Iranian export routes from the Caspian would be compelling. Nonetheless, Iranian behavior suggest its political evolution is likely to be gradual in regard to normalizing relations with the United States. China and Malaysia are already investing in Iran, and Japan is also actively pursuing investment in Iran's oil sector.

Geopolitically, the inability to qualitatively improve U.S.-Iranian relations may be inadvertently fostering tension with the Central Asian states it has sought to cultivate. These states, faced with necessity to export oil, have quietly contemplated a variety of oil and gas deals involving Iran, most notably the Turkmenistan pipeline. Iran is already tendering to build its own pipeline to bring Caspian oil shipped by tanker to a Caspian Sea port inland to refineries in northern Iran. Moreover, it is not improbable that the reformist parliamentary victory might strengthen Khatemi's hand in opening up Iran's economy more robustly to foreign investment in the energy sector. It has built at its own expense a natural gas pipeline that is transporting Turkmen gas to Iranian industries near the Iranian-Turkmen border. Iranian officials continue to pursue the possibility of a major export pipeline from the Caspian to the Persian Gulf.[29] U.S. chest-thumping may have delayed these projects slightly but is unlikely to stop them. Given the dominance of Soviet-style authoritarian and/or less than fully democratic leaders in Azerbaijan, Kazakhstan, and Turkmenistan, Iran may begin to look more stable and more pluralistic politically.

Not So Great Game

Socioeconomic problems threaten the stability of almost all of the governments of Central Asia and the Caucasus. This trend is likely to continue in lieu of the oil bonanza many in the region were expecting. The Caspian region's populations now spend on average 70 percent of their income on food alone.[30] In Azerbaijan, real GDP contracted by an average of 12.4 percent a year from 1992 to 1997.[31] Kazakhstan's GDP dropped 44.5 percent from 1990 to 1995, posting only minor recovery since.[32] As Martha Olcott notes in a 1998 *Foreign Policy* article, "Deep ethnic divisions have left each of these new states sitting on the equivalent of a separatist time bomb." On the deteriorating economic conditions she adds, "More troubling is the drastic decline in the ability of the Caspian governments to maintain even minimal levels of public services and social welfare protection, not to mention the kinds of benefits that the pre-independence population enjoyed. In the last several years, public education has broken down, health care has deteriorated, pensions have gone unpaid, and a relatively egalitarian social structure has been largely destroyed."[33]

Most of the ruling elite of the Caspian region have all but stamped out opposition movements and expression of political dissent, leaving no avenue for change except convulsive ones. Leadership successions can be expected in many of the region's states as the older generation of ex-Communists pass the reins of power. As a new generation of nationalists emerges to replace the old Soviet guard, social and ethnic tensions—as well as Islamic fundamentalism—may resurface in a more virulent form.

In the absence of substantial new energy discoveries in the Caspian region, the U.S. vision of Central Asia and the Caucasus as independent democracies buoyed by newfound oil wealth may yield to a much harsher reality. Similarly, the Asian vision of an alternative to dependency on Persian Gulf oil appears at best grossly inflated. By exaggerating the region's significance, the Clinton administration set in motion a policy bureaucracy and created a political investment that may be ill-advised. Indeed, burgeoning U.S. military and political involvement may invoke the law of unintended consequences. Ironically, this occurs as the private sector, with its eye on the bottom line, has had second thoughts about the Caspian.

Thus, a disconnect has evolved, where the private sector, which initially sought government help to gain entry to the region, has begun to drastically lower its expectations. Not only U.S. firms are more skeptical: China has indefinitely postponed a planned pipeline from Kazakhstan. Yet rather impervious bureaucracies continue on their merry way, failing to adjust to the new realities. Central Asia and the Caucasus are not about to spew forth an oil and gas bonanza. Nor must America brace for the threat to world stability that might erupt from a competition for influence there from surrounding powers. In regard to the latter, the experience of the past decade has been that relatively weak and inward-looking countries like Iran and Russia have been sucked into what might be dubbed a "Mediocre Game" for local spoils being largely played off by the various Central Asian actors. On balance, for the foreseeable future the principal threat to the region is and will likely remain internal instability, as with the Persian Gulf.

Nonetheless, the policy reality of the United States flies in the face of sensible and sustainable strategy. Promoting democratic evolution in the newly independent states is an admirable U.S. objective. But there are only very modest resources allocated to the sort of technical training and agricultural reform, humanitarian aid, and assistance in institution building necessary to achieve that goal. At best, the oil wealth to fund such development will be deferred for a good part of this decade.

But the impressive U.S. campaign for its Caspian basin objectives risks fomenting unnecessary strategic competition with Russia and China. U.S. policy focuses on expanding already obtuse military ties that constitute the principal interface between key Central Asian states and Washington. Indeed, in some cases direct support is lent to training military police and ground forces—in a policy reminiscent of ill-considered U.S. activities in Latin America in the 1970s. The U.S. European Command now has responsibility for the Caucasus—including a budding relationship with Georgia, and CENTCOM's portfolio includes all of the former Soviet Republics of Central Asia. The vehicle—and driver—for these liaisons is NATO's Partnership for Peace program, to which all the aforementioned states belong. Henry Kissinger has noted PFP is essentially therapeutic, aimed at psychological reassurance, "far removed from the basic NATO mission [and] which water[s] down the function of the alliance."[34]

S. Frederick Starr, in arguing that Central Asia is the keystone of Eurasian security, cites its geographic position as well as mineral deposits, oil, and stores of conventional weapons. He advocates that Uzbekistan is "uniquely positioned to anchor the security of the region."[35] Starr's position dovetails nicely with the views of some in the U.S. military. The United States has held joint military exercises in Kazakhstan and Uzbekistan and considers Uzbekistan (and Turkmenistan for that matter) an important listening post and bulwarks against Russian and Iranian expansion. The Uzbekistan exercises were designed to show off its ability to "land" inside the landlocked region. But airborne refueling might not have been as simple had a war situation really been in progress. There are no waterways for a U.S. aircraft carrier to float. The region is too remote for a quick flyover. Military involvement in Central Asia and the Caucasus would take land troops in substantial numbers. That would mean the U.S. people, Congress, and NATO would have to support it, and the stakes would have to be high enough to justify the cost in terms of human lives and material. It would also highlight competing U.S. policy goals of relations with Russia with those of supporting independence and democracy in the newly independent states of Central Asia. In short, the U.S. lacks the force projection capabilities as well as the vital national interest to intervene successfully on a sustained basis. Even in the process of redefining NATO's goals and missions, there would be extreme reluctance on the part of the EU countries to entertain the notion that NATO should be responsible for that peripheral area.

In any case, involvement in Central Asia and the Caucasus alone cannot significantly reduce the need for the United States to police the Persian Gulf, but such involvement could prove costly in political, military, and economic terms. For this reason, the United States would be better advised to encourage Russia into multilateral solutions toward the region's problems than to confront it with the option of taking over Russia's policeman role.

Russia faces a conundrum. Its own stability can be affected from either interfering too much or not interfering enough in the Caspian region's affairs. Respected Russian analyst Andrei Kortunov recommends his government acknowledge limitations and seek international assistance and cooperation. "Russia will be well advised to continue refining multilateral mechanisms that may assist it in the performance of onerous duties of conflict resolution in Central Asia and the Caucasus.... Additionally, it may try to attract greater international (U.N., OSCE, NATO, U.S., European) attention to mounting Central Asian problems to master additional support for peace-keeping activities in the region."[36] The idea that the United States could and should play a military role to "replace" Russia and guarantee the freedom of the Central Asian and Caucasus countries from the "Soviet grip" drips with impracticality. A careful glance at the international map demonstrates the huge costs and risks such a policy would entail.

These new Central Asian entanglements, quite apart from complicating already troubled U.S. relations with Russia and China, may create expectations that NATO will intervene in the event of social or political turmoil in places as remote and geopolitically significant as the border of China's Xinjiang province. While there are understandable reasons to be enthusiastic about the hope of fostering new independent, democratic states closer to Russia's flank, the vast entanglements such intervention might entail should not be considered lightly. The complex problems of the Caspian region—evidenced by the turmoil in Chechnya, Dagestan, and Kyrgyzstan—left to fester without the panacea of promised oil wealth could make the Balkan horrors pale by comparison.

As the promise of oil-driven nation building in the Caspian region proves elusive and underlying tensions produce full-fledged crisis, the United States stands a great risk of drifting into a quagmire that it can neither handle effectively nor solve militarily. As a Rand study noted, NATO risks a "mission creep and a developing gap between its commitments and capabilities." Rand concludes wisely that "NATO's military capabilities (read U.S.) . . . is of little utility in preventing 'failed states.'"[37]

Ironically, intense U.S. interest in the oil and gas of the region is sparking equally attentive focus from other national actors, not just Russia and Iran, who are necessarily part of the Caspian equation, but China and Japan as well. The Caspian region has begun to enter into both Beijing and Tokyo's energy security and strategic calculus in ways that may be unwarranted by the above-described realities of the Caspian. Both China and Japan have made significant investments in the region: China in Kazakh oil fields and promises of building pipelines; Japan in its aid and investment in the region (see chapter 7). Both are searching for diversity of supply. In addition, there is a geopolitical dimension to both Chinese and Japanese involvement in the region. For Japan, it seems aimed at limiting Chinese influence and defining its foreign policy as a major player in Eurasia.

In China, it appears that some analysts are mirror-imaging the U.S.-exaggerated importance of the region. For example, one prominent Chinese intellectual wrote an essay arguing, "The U.S. regards Central Asia as a component part of its global strategy and hopes that Central Asia will dissolve into the West's 'free market' and establish a western style 'democratic' political system. Central Asia's abundant natural resources and important strategic position are becoming more and more important in America's economic and security policy."[38] While one can find more benign and realistic views in Chinese policy journals and discussions with think-tank intellectuals, it is also conceivable that such misperceptions could grow into unnecessary strategic competition, or, given that China's Xinjiang province borders on Central Asian republics such as Kazakhstan, such perceptions of U.S. policy could feed Chinese anxiety about a U.S. containment policy.

China has immediate security concerns most notably as its volatile Xinjiang province, where ethnic Turkic Uighurs have exhibited serious political unrest. Moreover, conceptually, China views itself as a "landbridge" from Central Asia. More broadly, some Chinese planners view Central Asia as a means of diversifying and reducing dependence on Middle East oil. For Beijing, Central Asia, in this sense, offers the hope, however, unrealizable, of avoiding dependence on the U.S. Navy to safeguard its oil imports through the sea-lanes. This is a point to which geopolitical analysts and military planners in China have paid much attention.

Conclusion

Apart from oil—and U.S. firms dominate the consortia which already control the major oil fields of the Caspian—American national interests in the Caspian Basin are more derivative than fundamental. For China and Japan, they are also case-specific rather than broad-based and part of a wider focus related to stability in Russia, China, Turkey, and the Persian Gulf. They are a pale second compared to U.S.-Russian relations. On an international level, they are oriented toward ensuring that the region doesn't become a hotbed for illicit trafficking in arms, controlled technologies, and drugs, and a modest effort to facilitate moves toward political pluralism. Moreover, the lessons of the former Yugoslavia underscore the dangers of allowing rampant ethnic separatism in the Caucasus and Central Asia to stimulate political devolution in neighboring states such as Russia, China, or the subcontinent.

Apart from making a potentially costly policy mistake in Central Asia, the U.S. emphasis on the region appears to be playing a role in shaping Asian nations' views of Central Asia. Certainly, as we have seen, China's views of the Caspian region's strategic significance appear at least in part a response to what Beijing views as U.S. strategy. Rather then allowing for the possibility that a variety of single issue interests are driving U.S. policy in ways that may be contrary to U.S. larger strategic interests, there is a tendency to assume American behavior results from some "grand strategy." Similarly, Japan has followed the United States into the Caspian as well in a manner that suggests its perceptions are also shaped by American actions.

Finally, in regard to energy security for Asian actors as well as the United States, the key point is that the Persian Gulf's importance will not be mitigated by the energy resources of the Caspian. Rather, it will be modestly, but not unimportantly, supplemented. For Asia, the Gulf/Middle East will remain its principal supplier of oil as well as an increasingly important supplier of natural gas. The policy challenge for Asia and the United States is to rethink the Caspian and place it in the proper geopolitical and energy security perspective.

Chapter 4

The Asian Energy Predicament: A New Geopolitics?

Asia will be the world's largest consumer of primary energy by 2010, with the overwhelming majority of its oil imported from outside the region. By 2020, China and India—both with burgeoning middle classes—will produce more carbon emissions than the United States and Europe. Governments in Asia are more deeply involved in the energy sector than most OECD nations—from state-owned oil companies dominating all facets of energy production and distribution to price subsidies and regulating imports and exports. Historically, Asian governments have viewed energy largely in strategic terms, as one of the "commanding heights" of the economy and as a fundamental national security matter. Yet market forces (by spurring conservation, energy efficiency, new oil development and alternative fuel sources) rather than *dirigiste* policies have increasingly proven key to resolving energy questions. Asia is the only region of the world where the use of nuclear energy—with its attendant safety and proliferation risks—is growing significantly as a source of electricity. Moreover, this strategic view of energy, particularly oil, has been part and parcel of most Asian actors' national security and foreign policy calculus in a region where underlying suspicion, distrust, and rivalry remain part of the pathology of inter-Asian and trans-Pacific relations still in a state of protracted historic transition.

The question of energy in Asia, however, can be viewed through several different prisms. If viewed through the traditional, scarcity-tinged lens of "energy security," then what quickly emerges is the specter of a large and dynamic region increasingly dependent on a volatile area of the world—the Middle East—for upward of three-fourths or more of its oil imports. This is characterized in dire terms by one analyst as a "tightening embrace of necessity between East Asia and the Middle East that, over the next generation, could fundamentally challenge the Western-dominated global order."[1] In this view, Asia's thirst for oil is not only hastening the end of the oil age, but en route, it is increasing the likelihood of conflict over territorial disputes in the competition for increasingly scarce hydrocarbon resources. One variation of this perspective, discussed below, views these trends–Middle East producers needing Asian markets and Asians safeguarding secure supplies—as possibly coalescing into the West's worst nightmare: an "Islamic-Confucian" coalition. Such views tend to lead toward a subordination of market-based energy choices to perceived security imperatives, a geopolitics-centered rather than a geoeconomics-centered view of energy.

Viewed through the prism of market-driven approaches, however, a very different vision emerges, one of energy needs met through privatization, deregulation, regional integration, cooperative ventures, foreign investment, and absorbing technological innovation in a globalized, Information Age economy. This approach grasps the commodification of oil and gas markets, and the increasing role of technology in the energy sector (as discussed in the previous chapter). In this vision, as Daniel Yergin has argued, "stresses can be resolved not through massive armies and blue-water navies, but through markets and investment within the ever-denser web of international commerce."[2]

In this view, mobilizing the massive finance needed for energy infrastructures (estimated at over $1 trillion over the next decade) and commercializing new sources of energy (e.g., fuel cells, gas-to-liquids) will matter more than asserting claims for island real estate in whose territorial waters modest amounts of oil and/or gas may or may not exist. In this vision commerce and market forces shape inter-state relations more than ideological or nationalist imperatives shaping commerce and trade. This perspective sees the phenomenon of "globalization" not just as a source of volatility and instability, as evidenced in the Asian financial crisis. It certainly can be that. But globalization can also be a force for stability and prosperity, efficiently making sufficient quantities of resources available at lower prices, and pressuring domestic industries to produce and/or refine and market energy more competitively. Elements of both perspectives—and the policy choices they lead to—are competing in Asia, though the evidence suggests the region is gradually moving in the direction of market-centered trends.

In terms of Asia's future choices, the two perspectives on energy are not necessarily mutually exclusive, at least in the near-term. How Asian governments define energy security, however, will determine, to a considerable degree, whether the nations of the region make choices that take them in one direction or the other. Will Japan continue to pour resources into its plutonium reprocessing program, though it lacks any economic rationale in the chimerical hope of attaining future energy independence? Will India insist on continuing its risky, expensive nuclear power program, though natural gas is far more economical; or will Beijing sink billions into a pipeline from Kazakhstan in the hope of reducing dependence on Middle East oil? These are illustrative choices that would reveal which vision dominates Asian energy policies. Deeply ingrained Asian "scarcity" and statist habits suggest that it may be a decade or more before clear signs of major change are evident. At present there is what could be viewed as a third approach, with elements of both approaches co-mingling—at least for the near to medium term—as Asia, to borrow Deng Xiaopeng's description of China's economic reform strategy, crosses the river (of energy challenges) by feeling for stones.

Asia's phenomenal economic trajectory (prior to July 1997), its steadily rising energy needs (with 53 percent of the world's population), and its accumulating impact on world energy markets give the question of how the region meets its energy challenges a particular urgency and import.[3] The sheer magnitude of the issue is difficult to escape. Yet the Asian energy factor, snowballing during a period in the 1980s to mid-1990s when low oil prices removed energy as a frontline global concern, failed to garner much attention. It was Asia's growth that dazzled the world, not its attendant energy consequences. East Asia's role as an engine of global growth for nearly a quarter century was reflected in an average of roughly 6 percent annual GDP growth, a boom unprecedented in the modern era. By the mid-1980s, this began to generate the once familiar—now suspect—chatter about a coming "Pacific Century" and was dubbed by the World Bank in 1993 the "East Asian Miracle" in a much noted report.[4] As Asian growth rocketed from just 4 percent of world GDP in 1960 to roughly 25 percent by 1995, its energy consumption grew by similar proportions. By 1994, Asia-Pacific oil consumption rivaled that of the United States.[5] From 1971 to 1994, total primary commercial energy demand in East Asia (excluding China and Japan) grew by an average annual rate of 6.8 percent, quadrupling in absolute terms. In China, Korea, Thailand, and Indonesia, electricity demand grew even faster than GDP, averaging 11–12 percent in this period. In the decade 1983–93, four Asian economies—Japan, China, Taiwan, and South Korea—accounted for 36 percent of the world growth in primary energy demand.[6]

A symbolic turning point that began to put the spotlight on the Asian energy question was reached in 1993, when China, then the world's fifth largest oil producer, became a net oil importer. That led many analysts to begin pondering what the implications might be of another quarter century of the Asian miracle. Where would some 26 to 31 million barrels of oil (40 percent of total 1998 world production) that the Asia-Pacific was projected to need by 2020 (depending on moderate or high growth scenarios) come from, and how would such demand effect world markets, energy security, and not least, the environment? On the Asian side, how will the various actors respond to their respective plights of dependence on imports? Asia's economic weight in the world economy already was dramatically illustrated when, after being hit by financial crisis in mid-1997, the region's consumption fell to −2.7 percent in 1998 after averaging over 5.5 percent annual growth for the past two decades.[7] This development in 1998 led to a rare state of near zero-growth (0.5 percent) in world consumption of oil, as prices fell to pre-1973 levels, hitting $8 a barrel at one point, and left producers with an excess of more than a million barrels.

Yet for all the ramifications of—and intriguing questions raised by—East Asia's breakneck economic development achievements over the past quarter century, until the mid-1990s, energy questions—when they did get attention—tended to be largely confined to their own specialized intellectual or bureaucratic ghettos. With the possible exception of Japan, energy rarely occupied the center stage of debates over economics or security. Instead, they tended to focus on the marvels of export-oriented growth strategies and the virtues of the "Asian model." And why not? The Asian juggernaut appeared unstoppable—even through the oil shocks of 1979–80.

Similarly, until the mid-1990s, energy security did not figure prominently in the seemingly endless conferences and symposia on "the future of Asian Security" preoccupying many regional specialists. Instead, they focused on military modernization, flashpoints such as the Korean Peninsula, the Taiwan Strait, Kashmir, myriad territorial disputes, and by the mid-1990s, concern over the meaning of an emerging China. Apart from tensions over disputed (and allegedly, but unproven, oil and gas rich) islands in the South China Sea, energy, prior to the mid-1990s, was not often on the Asian security radar screen at all, despite the region's negative energy balance. Energy remained the domain of specialized economics and industry analysts. But gradually, the sheer volume of East Asia's real and projected energy needs—and their intersection with security issues—began to figure more prominently into the dialogue about the future of the Pacific. Thus, prominent regional specialists have suggested that energy is the "dark side to the explosive economic growth of East Asia, rooted in the region's profound energy insecurities."[8] This view, if not quite conventional wisdom, still holds considerable sway among many Asian security specialists, though most economists, and particularly energy specialists, dismiss such pessimism as unwarranted.

Presuming cool rationality as the basis of public policy, however (indeed, of any human behavior), can be a perilous act. That said, recent Asian experience points more in the sunnier direction of the economists than toward the darker view of the pessimists. Indeed, it is important to note here one of the most remarkable and under-appreciated facts about the Asian "miracle": all of the most dynamic and fastest growing "miracle" economies—Japan and the "four tigers" (South Korea, Taiwan, Singapore, and Hong Kong) possess virtually no hydrocarbon resources! Yet this energy deficit was simply not a significant factor affecting the region's most remarkable economic success stories. To this, add the instructive experience of the West in coping with the 1970s oil shocks in order to get a sense of how energy issues might be managed. If past is prologue, it should provide reassurance, or at least give pause to those anxious about perceived energy vulnerabilities when contemplating Asia's future energy choices and alternatives.

Whither the "Miracle"?

But the first question about Asia's energy future must be the impact of the 1997–98 Asian economic crisis on future economic growth, and hence, likely energy requirements and possibilities. Asian oil consumption decreased by some 500,000 barrels a day in 1998, though demand shot back up by more than that amount in 1999. One pessimistic (and unrealistic) IEA "low-growth" scenario suggested Asia's share of global demand growth may shrink to 23 percent by 2003.[9] Does the economic recession that still afflicts parts of East Asia signal the end of the Asian dynamism? Or was the contraction of 1997–98 more like a severe, protracted correction, signaling a new phase of the region's economic development? Economic historians will no doubt be debating the relative weight of various causal factors of what was arguably the worst financial crisis since World War II for some time to come. Nations that knew only robust growth for a generation suddenly faced deep recession in 1998: Thailand –9.4 percent, Indonesia, –13.4 percent (from 6 percent in 1997), South Korea –5.8 percent, Malaysia –6.7 percent, Hong Kong –5.1 percent, and Japan –2.8 percent.[10] Yet by fall 1999, all but Indonesia showed growth inching into the positive column (albeit, measured against precipitous decline) and at least tentative signs of recovery. South Korea, with a 10 percent growth for 1999, taking important steps towards restructuring its financial system and major changes in its business culture—not to mention rebuilding its foreign reserves to nearly $80 billion—was in the forefront of what is a still tentative recovery. (See Table 4.1.)

In retrospect, it appears the negative synergy of excessive capital flows, misallocated in the affected countries, fixed exchange rates, and overguaranteed, underregulated and opaque domestic financial markets were the key factors precipitating the crisis. These elements sparked a fourth—contagion in international financial markets—which caused the rapidity of onset and severity of the crisis.[11] Prior to the crisis, the region was a magnet for international lending, accumulating some $420 billion in net capital flows during the 1990s. Much of it was short-term money, lent out, too often imprudently, in local currencies. This led to lavish real estate development, overvalued property and stock markets, as well as industrial overcapacity. Wild swings in the dollar-yen relationship—the dollar rising 60 percent, and the Yen falling 60 percent—along with a 1994 Chinese devaluation wreaked havoc on economic competitiveness and heightened respective foreign debts. Devalued currencies meant that the price of oil imports (purchased in dollars) skyrocketed two or three times higher than before the crisis; affected Asian oil exporters such as Indonesia and Malaysia suffered declines in oil revenues.

These developments brought into sharp relief the fault lines of weak,

Growth Rates of Real GDP [1]
(percent per year)

COUNTRY	1981-90	1991-95	1996	1997	1998	1999	2000	2001	2002	2003	2004
Japan	4.3	1.5	5	1.4	-2.8	1	1.8	1.8	1.8	1.8	1.8
Singapore	6.3	8.4	6.9	7.8	1.5	5	6.7	6.4	6.2	6.1	6.5
Hong Kong	6.9	5	4.5	5.3	-5.1	-0.5	3	2.7	2.8	2.9	3
Taiwan	7.8	6.5	5.7	6.8	4.8	5.5	6.1	6.3	6.4	6	5.9
South Korea	10.7	7.5	7.1	5.5	-5.8	8	6.8	4.9	5.3	5	5.5
Malaysia	5.2	8.7	8.6	7.7	-7.5	2	5.7	5.2	5.5	5.7	5.9
Thailand	7.9	8.4	5.5	-1.3	-9.4	3	4.8	4.6	4.8	5.1	5.2
Philippines	1	2.2	5.8	5.2	-0.5	3	3.9	4	4.2	4.7	5
Indonesia	5.5	7.8	7.8	4.9	-13.2	2	4.3	4.1	5.1	4.6	4.7
China	10.4	11.8	9.6	8.8	7.8	6.8	7.2	7	7.2	7.4	4.9
India	5.7	5.2	7.8	5	6	7	6.8	6.5	7.1	6.4	7.2
Vietnam[2]	7.1	8.2	9.3	8.2	4.4	4	4.2	5.9	6.5	6.7	7.1

[1] Projected, EIU Country Forecasts (specific projections for Japan 2000-04 are an average estimate of Real GDP growth).
[2] 1991 Real GDP Growth Rate Unofficial IMF Estimate

Table 4.1 Growth Rates of Real GDP (percent per year)
Source: Economist Intelligence Unit and International Monetary Fund.

immature financial sectors and—with varying degrees of coziness—the "iron triangle" of government-finance-industry business cultures, hobbled by politically directed lending and unable to provide "early warning." These structural flaws had been camouflaged by massive input-driven growth (foreign lending, domestic savings). Like a swift kick in the stomach, the crisis deflated a cockiness about East Asia's economic prowess, replacing triumphalist rhetoric about the superior "Asian model" of capitalism with a new humility and modesty. Yet the surprising swiftness of recovery in much of the region also bears careful scrutiny. The private capital that fled amid the contagion has only just begun to trickle back in—dropping from $174.1 billion in 1996 to $8.6 billion in 1998, edging up to some $39 billion in 1999, with decline to $26.8 billion projected for 2000, as the crisis-driven interest of bargain-hunting foreign investors ebbs.[12]

Bad debt, both corporate and in the financial sector, in Thailand (bad loans at the end of 1999 accounted for 48 percent of total bank lending, 110 percent of GDP in Indonesia) and Korea has only begun to be worked out, and in Indonesia had barely moved forward at all by the end of 1999. Korea has made significant progress in addressing the weaknesses in its financial system and is at least beginning to change its business culture in a more market-based direction. Yet working out Daewoo's massive $73 billion debt, for example, will be a protracted process—but one that suggests change in how Korea's economy works, change toward a more market-driven system.[13] This is clearly the vision articulated by President Kim Dae Jung who has sought to erode economic nationalism and create a more open, globally competitive Korean economy.[14] And Japan, the world's second largest economy, remains flat, with fundamental but slow-motion restructuring of its corporate and financial sector unfolding amid declining property prices, and debt at the local and national level deepening (gross debt is $4.6 trillion, 110 percent of GDP). The bulk of Japan's meager growth has been propelled by massive government stimulus packages ($514 billion bank bailout, $288 billion to prop up small- and medium-size businesses in 1998–99); few analysts expect sustained Japanese growth to reach 2 percent before late 2001 or 2002.

Nonetheless, measured against numerous 1998 predictions of a three-to-five-year crisis, the region has appeared to stabilize and achieved relatively swift, albeit uneven, movement toward renewed sustained growth. To fathom why—and why sustained Asian growth will resume—it must be strongly emphasized that the fundamentals that led to the phenomenal Asian dynamism of the quarter century preceding the crisis remain in place. High savings rates, strong emphasis on education, hard work, good macroeconomic management, ability to absorb technology, and relatively open capital markets remain important assets that have not abated. These underlying characteristics and the adjustments made by the affected countries, combined with the response of international financial institutions, all help explain the turnaround. The delinking of Asian currencies to the dollar altered one source of volatility. Devalued currencies and cheap assets stimulated bargain hunters and kept foreign direct investment at roughly precrisis levels. The IMF commitment of $143 billion in emergency loans, particularly after the fund revised initial ill-suited conditionality of fiscal stringency, helped provide breathing space.

In this crisis-adjustment environment, stimulus packages initiated by Thailand, Korea, and other affected countries began to produce results: currencies and stock markets rebounded. Moreover, a process of steadily—if slowly—increasing transparency in the region's financial system began to unfold in 1998–99, including more serious bankruptcy laws, though still of varying efficacy. Ironically, the relatively quick return to economic growth will tend to slow, if not defer, needed economic restructuring by removing the sense of urgency for economic reforms that the crisis instilled. This, plus the continued debt overhang, has sparked debate about whether the signs of recovery are ephemeral or the beginning of well-grounded, sustainable growth.[15] Nonetheless, as a new century began, the Asian recovery was fragile and tentative, with sustainable growth across the region remaining vulnerable until remaining bad debt is restructured, banks recapitalized, and financial systems are modernized.

What all this adds up to is more modest and possibly erratic economic growth in much of the Asia-Pacific that may persist until 2002–2003. In terms of oil, by mid-2000, demand had picked up from the previous year's 4.8 percent in Japan, 11.1 percent in South Korea, and 11 percent in China.[16] Sifting through IMF, Asian Development Bank, revised IEA forecasts, EIA, Goldman Sachs and other private analysts' projections leads to some indicators of probable economic trends. Japan is likely to remain around 1 percent or lower, China unlikely to go above 6 percent growth before 2002, and growing debt from government stimulus packages and internal turmoil could scale back its economy further in the 2000–2003 period.[17] For the region as a whole, the most prudent forecast is for a moderate recovery in the near term, in the 4.5 percent to 5.2 percent range, reflected in the "moderate" recovery scenario of the IEA. In the 2004–2010 period, the higher end of that range is projected; while from 2011–2020, the lower end (+ or −.75) is a more likely outcome.[18] (See Table 4.1.)

Asian Energy Patterns

While the uncertainty factor (e.g., renewed recession, war, political turmoil in China, technological change) stalks all economic projections and cannot be discounted, even at what are likely to be far more modest growth rates than seen during the 1980s and early 1990s, oil demand—which generally lags, but tracks closely to growth rates—is still projected to grow substantially faster than the rest of the world. By 2020, according to the Energy Information Agency (EIA) of the U.S. Department of Energy, Asian oil demand will still grow two to three times faster than that of the industrialized West.[19] This assessment assumes oil prices *averaging* $15–$20 a barrel range during this period. It should be noted that such a demand growth projection is similar to the pattern of the previous quarter century (1970–94): Asian energy demand grew by 274 percent while that of the rest of the world grew at 63 percent during that period.[20]

While the economic crisis has altered the pace at which trends and patterns of energy use will unfold—and the time frame in which energy investment decisions are made and implemented—the region's recent reversal of fortune does not change the basic situation for any nation in the region. Nor does it alter the current and looming Asian energy factor in world markets: the region has a large and growing oil deficit, already some 12 million barrels, or roughly 60 percent of its oil is imported with limited capacity to expand production in the region. Already, China is the world's second largest consumer of energy, in 1997, India became the sixth largest. By 2020, the Asia-Pacific will consume more than one-third of the world's total energy; its oil imports will roughly double.[21]

Yet statistics can be deceiving in many ways. In the case of Asia, the aggregate numbers can be presented to conceal as much as they reveal about the region's energy patterns, given the distorting effect of the volume and energy mix of China. The more developed Asian economies—Japan, South Korea, Taiwan, Hong Kong, and Singapore—those whose energy demand growth will tend to be relatively more modest than others in the region over the coming two decades—account for only 7 percent of the Asian population but consume nearly half the region's oil. These more mature economies depend on oil for about 55 percent of their primary energy needs (Thailand and the Philippines also rely on crude oil for more than half of their energy needs), while in China, despite its pace-setting growth in oil demand, oil only accounts for 22 percent of its energy consumption. China, dependent on coal for 68 percent of its energy needs (India depends on coal for nearly 60 percent of its primary energy), also skews the fuel mix for the region, for which coal accounts for 46 percent of total energy use. Similarly, the region as a whole uses slightly less oil (38 percent of total energy) than the world average; substantially less natural gas (8 percent) and less hydropower (1.4 percent).[22] (See Figure 4.1.) This energy mix, with coal and oil providing 84 percent of Asia's energy, explains why the region is a leading source of carbon emissions.

More remarkable than the dimensions or even the growth of Asia-Pacific energy consumption is the fact that these energy realities occur at the same time that Asian per capita energy consumption is barely half the world average! The United States, for example, consumes eleven times more energy per capita than China, and nearly twenty times more per capita than India. But the more developed economies of the region—Japan and the "Four Tigers" (South Korea, Taiwan, Singapore, Hong Kong) have per capita energy consumption levels closer to those of Europe and the United States. At the same time, the level of energy efficiency, measured in energy use per unit of gross domestic product (GDP) is the inverse. China, for example, requires roughly four times the oil equivalent in energy consumption per thousand dollars of GDP of the United States, and nearly ten times that of Japan, an unusually energy-efficient economy.[23] Moreover, China's per capita oil consumption is nearly 22 times less than that of the United States, and 13 times lower than that of South Korea. Yet China's per

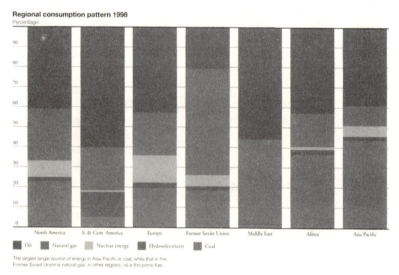

Figure 4.1 Regional Consumption Pattern, 1998
Source: BP Statistical Review of World Energy, 1999.

capita oil consumption is nearly twice that of India's. The differences are of similar proportion in regard to electricity consumption, with per capita usage in China about 5 percent of the OECD average, and in India just over 3 percent.[24] (See Figure 4.2.)

These patterns reflect respective levels of development and complexity of energy use. As evidenced above, advanced industrial economies with higher living standards tend to have much higher energy use per capita than developing nations but also tend to be relatively static in their consumption levels (which may even decline slightly). Energy intensity—the amount of energy needed per unit of GDP—tends to diminish in postindustrial societies that have moved to more services and knowledge-based economies. In contrast, developing countries rapidly industrializing and urbanizing with a large agricultural base (e.g., China, India, Thailand) tend to have large noncommercial energy use (e.g., biomass) but move to a higher energy intensity, and toward oil products as they modernize agriculture and industrialize. This also includes different mixes of oil products (e.g., from kerosene to fuel oil). Much of this demand growth tends to come from the transport and residential use sectors as a country takes off and a large urban middle class is formed. In Asia, this has translated into

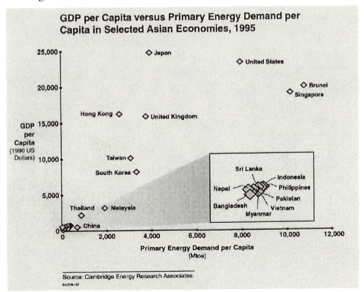

Figure 4.2 GDP per Capita versus Primary Energy Demand per Capita in Selected Asian Economies, 1995
Source: Cambridge Energy Research Associates.

rapid expansion of oil demand and projections of even higher increases in electricity demand with frequent fears of brownouts disrupting industry and private life.

The truly breathtaking experience of Korea, with energy demand tripling from 1975 to 1992, and oil demand quadrupling in the decade 1985–95, may not be replicated in nations much larger, more rural, and more segmented in terms of development such as India and China. However, the miniscule per capita consumption figures of both India and China—even compared to other Asian economies—foreshadow qualitative leaps in energy consumption over the coming generation. The recent experience of several Southeast Asian states may be instructive. A phenomenon evident in dynamic ASEAN economies (e.g., Thailand, Malaysia, Indonesia) less developed and economically complex than Korea was that fledgling middle classes, clustered in major metropolitan areas—in the case of Thailand, largely around greater metropolitan Bangkok—rapidly moved to obtain motorized transport, as incomes roughly doubled. In Thailand, for example, personal automobile purchases increased by nearly 18 percent annually from 1985 to 1992, and prior to the 1997, Thailand and other booming economies in Southeast Asia were increasing automobile purchases by nearly 30 percent annually. Yet each of these economies have significantly less than half the autos per capita of Japan.[25]

In terms of probable growth in energy demand in the first two decades of the twenty-first century, such consumption trends allow one to begin to grasp the dimensions of demand growth in larger developing countries, particularly, China and India, and if Southeast Asia is taken together, a group of some 500 million people. For illustrative purposes, China, which accounts for roughly half of Asia's energy equation, offers a useful glimpse of the potential for future demand growth. Until the mid-1980s, few Chinese had more appliances than televisions. Now, most urban Chinese have televisions (about 280 million), stereos (about 300 million), and fans, and about half, refrigerators and washing machines. Since 1978, the total number of air-conditioning units has grown more than fifty-fold.[26] During the 1984–96 period, residential consumption of electricity more than quadrupled. Yet China's per capita electricity consumption is only 8 percent of the OECD average.[27] And the next item on the wish list of many Chinese families is their own automobile.

Multiply this phenomenon of an emerging middle class, already in the case of China, roughly 100 million, by the perhaps 200 million more (of the 1.5 billion) in India and Southeast Asia, and the proportions of the consumer revolution in developing Asian states begins to register. Already, it is shifting the balance of oil consumption in the region. In the decade from 1988–98, Japan's share of total Asia-Pacific oil consumption shrunk from about 40 percent to 28 percent.[28] In regard to East Asia, a more economically integrated unit, Japan's share of oil imports has begun to sharply decline, from 77 percent in 1992 to about 56 percent in 1998, and is projected to decline to 36 percent of East Asian imports by 2010, while the imports of Greater China (China, Hong Kong, Taiwan) are projected to reach 28 percent by 2010.[29]

Asian Supply and Demand

The distribution of Asian energy demand, particularly, oil demand will continue to shift toward China, India, and Southeast Asia. But Asia-Pacific consumption will continue to expand rapidly, if at a somewhat less frenetic pace than pre-1997 growth rates. Already, Asia has a substantial oil deficit. In 1998, the Asia-Pacific imported 11.5 million barrels a day of the 19.1 million barrels it consumed daily, with imports rising to 12 million b/pd in 1999. This amounts to about 62 percent of total petroleum products consumed. Asia-Pacific import needs are projected to approach 17–19 million b/pd by 2010, and the EIA forecasts those import needs to rise to 24 million b/pd by 2020. China's import needs alone are projected to grow to roughly 3 million b/pd by 2010, and as much as 5–6 million b/pd by 2020.[30]

The composition of Asian energy demand, however, is unlikely to change more than incrementally over the next decade. One important caveat, however, is growing environmental concern, the wild card in energy decision making. Pressures resulting from concerns about the human contribution to global warming and industrial pollution could alter energy consumption patterns more rapidly than "business as usual" projections suggest. For example, apart from horrendous pollution in major cities, acid rain from use of coal, particularly in the "rust belt" in Manchuria, has become a problem across Northeast Asia. Nonetheless, the large coal reserves in China and India make it likely that coal will remain disproportionately large in their respective energy equations. Problems of public acceptance as well as management of nuclear waste are likely to limit the growth of nuclear energy in Asia, Japan, South Korea, and Taiwan. In the case of China, the larger impediment to planned expansion of nuclear energy is principally the cost. But the effect is the same: major scaling-back of nuclear energy plans over the coming decade.[31] Even the most optimistic projections of China expanding the use of natural gas, hydropower, and/or nuclear energy to displace coal in its energy mix envision a reduction of its reliance on coal to no lower than 60 percent by 2020.[32]

In any case, natural gas, which has been steadily growing at almost 18 percent annually, still only comprises less than 10 percent of Asia-Pacific primary energy consumption, far below OECD levels of about 26 percent. There is also large potential for the expansion of Asian production of gas, with Indonesia, Malaysia, and Brunei already major exporters. Natural gas resources in China are underexplored and are likely to prove substantially larger than the current 48 trillion cu. ft. (more than half the proven reserves of Indonesia) proven reserves.[33] Moreover, there are massive natural gas resources in the Russian Far East, roughly equivalent to world gas reserves to what Saudi Arabia's oil resources are to world oil reserves (discussed in chapters 7 and 9). There are several pipeline schemes that have been under discussion in the region, particularly one seeking to link Russian gas in Irkustk into China and Northeast Asia in an energy grid. But this massive infrastructure project is at least a decade or two away from being fully realized. Because it is plentiful and a relatively clean fuel, natural gas is fast becoming the fuel of choice for much of the world. There is also a large potential for substantial consumption of gas in Asia, which is already starting to be realized. In Asia, expansion of both liquified natural gas (LNG), currently imported largely by Japan, South Korea, and Taiwan, and pipeline gas, from within the region, of which there is currently very little, will gradually increase. As both require substantial infrastructure investment, it will be a gradual process over the coming two decades and will tend to substitute for coal and oil-fueled power generation. Gas will be increasingly important in avoiding one of the key strategic bottlenecks to economic growth—critical electricity demand in the region—as well as for residential use, but building the required infrastructure will be a protracted process over the next quarter century.[34] It must also be noted that Indonesia and Australia have large reserves of low-sulfur coal, seeking export markets. In light of the growing environmental factor in energy choices, such a relatively clean energy source available within the region is likely to compete with other fuels at least over the coming decade.

About half of the growth in world oil demand over the next two decades, according to the EIA, will come from the transportation sector, and in the Asia-Pacific, the transport sector may well comprise an even larger portion of regional demand growth.[35] This raises the question of how much oil is likely to come from regional producers. Specifically, has Asian oil production peaked, or is it approaching its peak, or is there more room for growth? Many forecasters appear to have been unduly pessimistic about the prospects of Asian oil production. In 1998, Asia–Pacific production was 7.65 million b/pd, just over 10 percent of world production, but comprised just under 5 percent of world reserves. But the reserve/production ratio (how much proven oil remains if production continued at current rates) of Asian oil states is 18 years, relatively low measured against the 43–year world average reserve.[36] China, producing 3.2 million b/pd, accounts for about 40 percent of Asian oil production, with Indonesia producing half that of China, 1.6 million b/pd. India, Malaysia, Brunei, Vietnam, and Papua New Guinea account for the rest of the region's production. Indonesia, Malaysia, Brunei, and Vietnam are the only net oil exporters in the region, and some forecasts suggest that they may be net importers by 2010.[37]

There is reason to believe, however, that most forecasts that Asian production is near its peak and will begin to decline over the next decade may be overly pessimistic. This is not because of anticipated new finds of major fields. Indeed, the recent experience of oil exploration efforts in the East and South China Seas, and in western China's Tarim Basin—the most likely venues of new finds—has left Western oil firms rather skeptical about prospects of any such major commercial finds.[38] Rather, the reason pessimists may be wrong is in underestimating the potential of existing oil fields. A sampling of four prominent analysts brought forecasts ranging

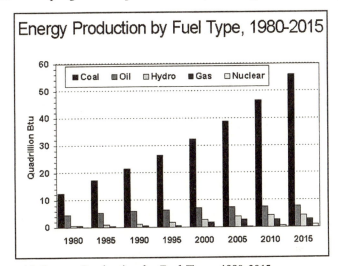

Figure 4.3 Energy Production by Fuel Type, 1980–2015
Source: U.S. Energy Information Administration, "Country Energy Profile: China," October 1997.

from declines of 1.1 million b/pd to one of essentially maintaining current levels.[39] Comparing respective growth in Asian reserves to those of the North Sea from 1975 to 1995, Paul Hornswell makes a cogent case that, just as the North Sea oil defied most predictions, Asian regional production may be able to rise to 9 million b/pd or higher by 2010. The obstacles to enhancing production in existing Asian oil fields, he argues, are principally government policies, not geology. This view was also conveyed to the author by a number of Western oil company officials.[40] Achieving higher production may be possible if Asian producers remove the impediments to enhanced investment in technologies (e.g., 3–D and 4–D seismic and directional drilling) that can improve and prolong output. The principal impediments are seen as fiscal regimes (complex price setting and taxation) of various governments, particularly Indonesia, the most stagnant producer, and Vietnam, and inefficient industrial structures that could be liberalized to offer more incentives for investment.

The Asian-Middle East Energy Nexus

It was a sign of the times—virtually unnoticed in the U.S. press—when President Jiang Zemin made the first ever visit by a Chinese head of state to Saudi Arabia in November 1999, heading to Damman, the heart of Saudi oil country after a stop in Riyadh. That China, one of the world's largest oil producers, as well as other Asian countries are diligently cultivating energy ties to the Saudis and other major Gulf producers underscores an awareness of the limits of Asian oil production. For whatever the trajectory of Asian production—presuming no new major finds—it will, in aggregate terms, only affect the Asia-Pacific regional oil deficit at the margins, though unexpected enhancement of current production could be important to producers, such as Indonesia, Malaysia, and Vietnam.

The dominant, and seemingly inexorable trend is one of growing Asian oil dependence on the Middle East, and vice-versa. Already, over 60 percent of Middle East oil exports go to Asia, and nearly 70 percent of Asian oil imports come from Middle East producers. Moreover, the oil throughput for refineries in Singapore, Japan and South Korea, three major Asian refining centers, is overwhelmingly from the Middle East: Singapore (84 percent) and Japan and South Korea (78 percent). New refineries elsewhere in the region have also begun to look to the Gulf/Middle East in making decisions about crude oil purchases. These trends are relatively recent. From 1975 to the early 1990s, OPEC Middle East exports to East Asia grew only by about 40 percent. This was the result of several factors: Middle East production cutbacks to maintain high prices in the early 1980s; Asian reduction in demand in response to the two oil shocks by diversifying away from oil, especially in the case of Japan; increased efficiency of industrialized Asian consumers; and growth in Asian production.[41] (See Figure 4.4.)

But the current burgeoning Middle East–Asian oil nexus that began to take shape in the 1990s appears a permanent structural feature of the new global, Information Age oil market that has changed dramatically over the past two decades. Gone are the days of exclusive, secretive fixed contracts. As discussed in chapter 2, oil products have become fully commodified, mainly sold on a free market, floating price basis. International crude oil markets have become increasingly global and transparent, similar to the now familiar world financial markets that shuffle $1.5 trillion electronically around the world each day. In this new, more efficient oil market, oil export movements are driven in large measure by transportation economics, as most sellers seek to maximize revenues by finding proximate end users.

As oil commerce has come to be shaped more by transport costs than political relationships, the result, as discussed above, has been a largely bifurcated global market, with oil flows from the Middle East gravitating to Asia; while oil supplies from the Western Hemisphere (Mexico, Venezuela, Colombia, Canada) and the Atlantic Basin (North Sea, West Africa) have to a large degree displaced Gulf oil from the U.S. market. According to oil economist Fadhil Chalabi, a former senior OPEC official, by 2010, "the share of the Middle East's oil exports going to the Asia-Pacific region as a whole is expected to rise . . . to 66 percent. . . . Interdependence between the Asia-Pacific and Gulf regions in the oil sector is firmly entrenched."[42] Indeed, the Asia-Pacific will likely be importing 20–24 million b/pd from the Middle East by 2020, 95 percent of its total oil imports by 2010.[43] There may be some measure of uncertainty over the precise volume—whether Asian production increases, economic slowdowns weaken demand, or new technologies reduce oil demand (the latter may be a significant factor in the out years from 2015). But even if such unanticipated developments reduce Asian oil imports by one-third, the basic phenomenon of Persian Gulf/Middle East–Asian oil mutual dependence remains, whether it is 16 million b/pd or 24 million b/pd.

Yet if Gulf/Middle East–Asian energy interdependence is an emerging fact of life, its wider significance is a matter of much speculation and debate. As of mid-2000, the U.S. intelligence community has not done a major intelligence estimate on the security implications of this long-term trend. The cosmic question is what this phenomenon of an energy linkage

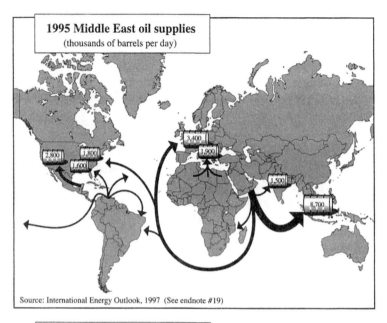
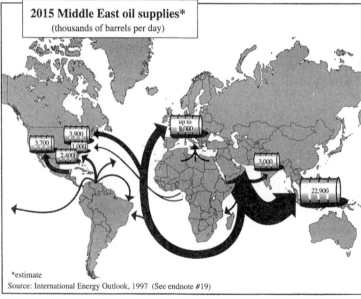

Figure 4.4 Middle East Oil Supplies, 1995 and 2015
Source: Geoffrey Kemp, "Energy Superbowl: Strategic Politics and the Persian Gulf and Caspian Basin," (Washington, D.C.: The Nixon Center, 1997).

between two potentially turbulent regions, the Gulf/Middle East and the Asia-Pacific, may mean in regard to the international system of relations. Is it principally a dynamic, complementary commercial relationship, or could it "fundamentally challenge the prevailing Western-dominated political order," as one prominent analyst has suggested?[44] Even if one views it less apocalyptically, what are the geopolitical implications of it for Persian Gulf security, for Asian security, and for the U.S. role in both Southwest and East Asia? The implications of the Middle East–Asian oil nexus may be initially divided into three related categories, the near-term (to 2010) and tangible, the long-term and intangible, and the potential nightmare scenarios. In short, the good, the bad, and the ugly.

Which Future: Geoeconomics or Geopolitics?

Both sides of the new oil equation are keenly aware of their respective predicaments, and have become increasingly animated by their respective imperatives: the need to ensure customers, on the part of Gulf/Middle East exporters; and the need to ensure oil supplies on the Asian side. "For the first time we are focusing on Asia," former Saudi Oil Minister and senior OPEC official Ahmed Zaki Yamani, reflecting fears at the time of the impact of the Asian crisis on OPEC, said in a keynote address to an annual London energy conference in 1998. "Asia," Yamani explained, "can play a crucial role in helping to improve OPEC's prospects."[45]

Thus, one central consequence of the Middle East–Asia oil link that has begun to unfold is a complex set of interlocking economic and financial relationships, a deepening commercial network more closely weaving the two regions together: Asians investing in upstream oil and gas sectors in the Middle East, while OPEC Middle East exporters invest in downstream Asian activities. As OPEC secretary general Rilwanu Lukman explained in a July 1998 speech at the Middle East Institute in Washington, "Asian investors, traders, operators and others are active in the energy sectors of the Gulf. On the other hand, Middle East companies and individuals have significant interests in the downstream energy industry in Asia."[46]

One of the more visible manifestations of this active but low-profile trend was the aforementioned unprecedented Jiang Zemin visit to Saudi Arabia in November 1998, during which he proclaimed that China is forming "a strategic oil partnership" with Riyadh. This was an intriguing choice of words, considering Beijing's penchant for declaring more generic "strategic partnerships" with Russia, South Korea, France, and other countries. In 1995, Beijing decided to import 3.5 million tons of crude oil from

Saudi Arabia annually (less than 1 percent of Saudi production). Jiang brought an entourage of Chinese business officials, signed oil cooperation agreements, and discussed a $1.5 billion refinery and petrochemical complex in Southern China that would use Saudi oil.[47]

There has been an accelerating pace of exploring and in many cases, concluding business deals between Asian and Gulf/Middle East officials in recent years. In addition to its efforts in Saudi Arabia, China has invested in oil fields in Iraq and Iran since the mid-1990s. Similarly Malaysia's state-run Petronas is in a joint venture with European companies to develop the South Pars gas fields in Iran as well as oil fields in Iran and Yemen. Both India and Indonesia have explored investing in the Tuba oil field in Iraq. Japan has steadily cultivated Gulf producers, with investment in Saudi Arabia, in Abu Dhabi, and in oil and gas in Oman, and along with South Korea has invested in Qatar to produce LNG. Japan developed the Khafji oil field in the Neutral Zone, divided between Saudi Arabia and Kuwait in the 1960s. The Khafji field produces 300,000 b/pd; 70 percent goes to Japan, part of nearly 1 million b/pd Japan imports from Saudi Arabia. But in early 2000, Japan failed to renew its production contract to the Saudi field.

In the other direction, OPEC Middle East producers have similarly been actively seeking to deepen their economic ties to Asia, with an emphasis on downstream activities, such as refining. The Saudis have been perhaps the most active in this area over the past decade, with the Saudi press covering visits of senior Saudi officials to China and other Asian nations with great fanfare. In 1991, Saudi Aramco bought a 35 percent stake in Ssangyong Oil Refining Co., the third largest refiner in South Korea, for $470 million and may increase its stake. Saudi Aramco also purchased a 40 percent stake in Petron, a major Philippines refiner in 1995, and in 1997, began a joint venture with Exxon to expand a refinery in China's Fujian province from 80,000 b/pd capacity to 240,000 b/pd. In addition, at the same time, the state firm also formed an alliance with Royal/Dutch Shell to invest in Asian downstream ventures. Kuwait and Qatar have also cultivated Asian importers, in the case of Qatar, seeking to lock in long-term gas contracts with India, China, and Japan. China has also been strengthening economic ties to Iran including agreements to increase oil imports and to build a joint refinery in China, as well as its oil investments in Iran.[48]

It is important to note that this phenomenon of interlocking energy investments is only in an early, and to some extent, exploratory phase. But it appears an imperative, reflecting underlying anxieties about energy security on both sides. As the oil and gas relationship with Asia expands over the coming decade, there will be an enormous volume of capital flows into the OPEC Middle East producers. According to one estimate, by 2010, if Asia is importing 17 million b/pd from the Middle East, at $20 a barrel, the result would be capital transfers to the Middle East of $124 billion annually.[49] Even in today's global financial markets, where nearly $2 trillion a day floats through cyberspace, that is real money. Such revenues could in part be recycled into downstream investment in dynamic Asian economies. But there are numerous other possibilities. Growing capital flows to the Middle East would go some distance in ameliorating a growing list of problems in major oil exporting countries such as Saudi Arabia, Kuwait, Iran, and Iraq. Riyadh, for example has seen a significant decline in living standards over the past fifteen years, has a large demographic bulge of youth to absorb into its economy, and has accumulated $130 billion foreign debt. But such capital flows could also accelerate efforts at obtaining a new cycle of modern weapons, including weapons of mass destruction.

If at least the rough outlines of the anticipated economic and financial consequences of the Middle East–Asian energy relationship for the global economy are discernible in terms of orders of magnitude and economic patterns, the political/security implications enter the realm of the intangible and the speculative. In the 1930s, it was energy security that led Japan to occupy Indonesia and control its oil fields. Indeed, the U.S. oil embargo was an important factor leading Tokyo to attack Pearl Harbor, which brought the United States directly into World War II. Some analysts see China, as a rising power with a newfound energy dependence, as a possible latter-day version of that experience. "The problem for Asian stability, growing with each barrel of Chinese oil imports, is now clear," writes Kent Calder in an influential book on energy and security in Asia. "It is the danger," he argues, "that China's attempts to safeguard its oil supply lanes and defend its historical sovereignty in adjacent seas poses for other nations, especially Japan. China claims 80 percent of the South China Sea as territorial water, 70 percent of Japan's oil supplies pass that way." Thus, this chain of logic goes, "As Chinese imports steadily rise, defending the fragile sealanes to the far-off Persian Gulf becomes a new security imperative for the PLA Navy." This begins to move from the second category of intangible into the third, of nightmare scenarios: Chinese destroyers and aircraft carriers disrupting tanker traffic in a war over disputed islands in the South China Sea; or in a war with Japan over the virtually uninhabited Senkaku islands; or worse still, allying with Iraq or Iran in a future Gulf war.

To be sure, the number of oil tankers plying the waters of the Indian Ocean, through the straits of Malacca, and the South China Seas for ports in Pusan, Yokohama, and Shanghai in the two decades ahead will likely increase threefold. But is there indeed an ominous security threat, as some fear? Again, it depends to a considerable degree, which lens energy security is viewed through. The scarcity notion can easily lead to imports equals shortages equals resource competition equals territorial conflict. The "looming conflict" model seems to be based on a sort of neo-mercantilist set of assumptions reminiscent of nineteenth-century and pre-World War II interstate relations in Europe. Certainly, the Asia-Pacific has more than its share of bristling nationalism, historical grievance, and ethnic antagonism. Yet thus far it has not tended to manifest itself in a manner similar to the experience in Europe in the century leading to World War II. But it must be noted that sealane security has become part of the mission of the PLA Navy. It is also much discussed by Indian defense specialists, and is a growing concern of Japan.

But there does not appear to be any imminent shortage of oil; nor, even if high-end demand projections are realized, is a shortage likely over the next quarter century. A study by the Baker Institute at Rice University based on median assumptions of a cross section of prominent industry consultant groups concluded that even assuming high-end assumptions of Chinese demand of 7.1 million b/pd by 2010, if non-OPEC production grew at 1 percent, a conservative forecast, the result would be a modest oil surplus of oil and prices under $20 a barrel.[50] Indeed, the Saudis and other Gulf states could fill most of the increased import needs of Asia-Pacific nations. And unlike the situation during the 1930s, foreign investment and long-term contracts are more efficient ways to safeguard supply than real estate grabs, and more likely to be the arena for resource competition.

China has not had a blue water navy for nearly six centuries, since the days of Admiral Zhang He during the Ming Dynasty. But neither has it been heavily dependent on foreign energy sources for the past six centuries either. In any case, while Chinese military modernization continues apace, it is unlikely that China will obtain the air and sea force projection capabilities to challenge the United States in the Gulf or in the Pacific for at least the next two decades. Nonetheless, there have already been rumblings about China seeking aircraft carriers, and recent announcements of Beijing's attempt to create a long-range air force, as well, certainly indicate the ambition to eventually create a blue water navy, even if it remains a generation over the horizon.[51]

Moreover, the Persian Gulf is not an area where China has had historic ties or compelling strategic interests apart from its general posture of countering "hegemony"—Soviet influence during the Cold War, and to some degree, the preeminent American superpower now. Yet the Middle East is fast becoming a significant trading partner for China, though Chinese exports to the entire Middle East—even after growing exponentially—are less than $5 billion. Beginning with the Iran-Iraq war, China developed an arms supplier role in the region, which continued into the 1990s, particularly in the form of missiles and other weapons of mass destruction to Iran. In one blatant 1997 episode, Beijing threatened to withhold support for continuing sanctions against Iraq if Kuwait did not agree to a major arms purchase, of self-propelled howitzers worth about $300 million.[52] At the same time, Beijing has developed a flourishing relationship with Israel and has supported the Middle East peace process. As was evident in its behavior during the 1990–91 Gulf War, when Beijing went along with U.S.-led efforts to reverse Iraq's invasion of Kuwait and with subsequent sanctions, China's top priority of economic modernization has been a dominant factor.[53]

But interestingly, as the specter of Middle East oil dependency has grown, China's arms sales to the region have tapered off, particularly to its most

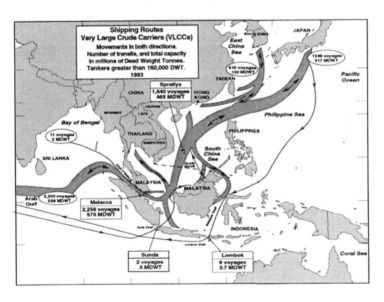

Figure 4.5 Oil Traffic and Sealanes (Subtitle: Supertanker Movements: Millions of Dead Weight Tonnes (MDWT)
Source: John H. Noer and David Gregory, "Chokepoints: Maritime Economic Concerns in Southeast Asia," Center for Naval Analysis/Institute for National Strategic Studies, (Washington, DC: National Defense University Press, 1996).

troubling customer, Iran, which Beijing had sold C802 missiles, other sophisticated military equipment, and planned to build two nuclear plants and a hexaflouride plant that could facilitate Iranian ambitions of joining the nuclear club. Some analysts suggest this has occurred because economically troubled Iran was in arrears to China for some $900 million.[54] Another factor may have been Beijing's desire to improve relations with the United States. But it also may be the case that China has reconsidered its dealings with Gulf/Middle East nations in light of its growing dependence on oil from the region. While China's behavior retains some ambiguity—selling technology and expertise rather than whole weapons systems, or in some cases all but complete systems—the trajectory of China's policies on missile and nuclear proliferation has been generally in the direction of international norms—the NPT, the Chemical Weapons Convention, the Comprehensive Test Ban Treaty (CTBT), and pledges to adhere to Missile Technology Control Regime (MTCR) standards.[55]

Certainly, highly dependent Asian energy importers are likely to have a new appreciation for stability in the Persian Gulf. This can manifest itself in a variety of ways, one of which might be the development of a new sense of common interests with the United States in the Gulf. This is a point U.S. officials have repeatedly made to China in an effort to persuade Beijing that missile sales to Iran, for example, may not be in China's best interest. The geoeconomic argument views oil as a globalized commodity. Thus, a disruption anywhere is a price spike everywhere. The competition for resources is most likely one of determining the best price—who gets what long-term contract—rather than access to the strategic resource. There may well be problems of short-term disruption, as occurred during the Iranian revolution in 1979. But turbulence in Saudi Arabia or elsewhere in the Gulf would likely mean only short-term interference and price hikes, not long-term competition for access to resources.

Even so, that still leaves China and other Asian nations free-riders as the United States remains the guardian of stability in the Persian Gulf. It is a legitimate question whether Beijing is comfortable remaining in such a position. Arguably, given the alternative of naval arms races, it might be argued that Washington may want to encourage China's free-riding. In any case, China and India both envision building more capable maritime forces over the coming generation. How much of a premium either nation may place on maritime control of sea-lanes far beyond the South China Sea, or the Indian Ocean, respectively, remains to be seen. But disrupting the flow of oil tanker traffic would cause equal-opportunity damage, driving up the world price of oil. It is a clear *casus belli,* the economic equivalent of using nuclear weapons. Given its dependence on Middle East imports, barring extreme circumstances such as World War III, it is difficult to see how China would benefit by interdicting oil shipping. Similarly, it is difficult to envision circumstances under which the United States would interdict oil shipments to Asia short of a protracted global conflagration. Regardless, issues of nationalism and sovereignty rather than energy security may be more operative factors animating Asia-Pacific political/military decisions.

Even so, there are two potential political questions raised by the Middle East/Asia energy connection that must be asked. One is the degree to which these burgeoning commercial relations acquire political weight. For example, during the 1990–91 Gulf war, China abstained when the United States mobilized a coalition to reverse Saddam Hussein's invasion and occupation of Kuwait. As Geoffrey Kemp has argued, if there is a future crisis in the region in which China or India, for instance, must choose between supporting directly or indirectly U.S. actions that involve oil or gas suppliers to either Beijing or Delhi, the international political results might be very different.[56] More broadly, will the collective economic weight and shared interests of the Middle East–Asia economic nexus take on a political expression, if for example, there is a growing perception of the United States as a global bully, acting unilaterally to coerce or shape behavior that either or both do not view as being in their respective national interests? Over time, this emerging Asia-Middle East nexus could erode U.S. global influence.

The second political issue is that of burden-sharing. For the forseeable future, the United States will remain a guarantor of stability in the Gulf. There is simply no other likely alternative on the horizon. Yet the resources the United States is protecting increasingly go elsewhere, largely to Asia, secondarily to Europe, but only marginally to the United States. It can be argued that since oil is fungible the United States does benefit from its role in the Gulf in any case. Yet there still remains the burden-sharing argument. U.S. allies in Europe lack the capacity, and to some degree, the political will to assume more political-military responsibility. Japan is constitutionally prohibited from assuming a proactive military role, and certainly for the forseeable future, not likely to have a Japanese (not to mention regional) mandate to project force at great distance from the home islands. Will the American public and their elected representatives continue to support such a U.S. military role in the region even if those directly benefiting from the U.S. stabilizing role do not assume their fair share of responsibility?

Unfortunately, these necessarily remain unanswered questions. How the answers unfold will depend in large measure on how Asia-Pacific nations conceive of energy security, how the political environments both in East Asia and in the Gulf/Middle East evolve, and what the energy picture looks like a generation hence. Whether the uneven process of globalization trumps historic animosities in the Middle East and historic fears in Asia is an open question. It is the interaction between the two that is likely to determine the outcome to the larger impact of the Middle East–Asia energy linkage on both region's security. This emerging relationship between the two most prominent non-Western civilizations on the Eurasian landmass, a central fact in world energy markets, is emblematic of the historic shift in the center of gravity of the global economy toward the Pacific Rim that is certain to be one hallmark of the twenty-first century. It is a phenomenon that bears careful scrutiny over the next generation.

Chapter 5

China: In Search of an Energy Policy

It is difficult to conceive of anything as emblematic of the challenges facing China's historic transformation—or of the work-in-progress character of its reforms—than the enormous and complex question of energy. For all the sound and fury of late about the rise of China, it is the growing impact of China's energy needs on global energy markets, the environment, and international security that perhaps best illuminates the consequences of an emerging, modernized China for the world system. Some might argue that as much as its military modernization, its next generation ballistic missiles, and its nuclear weapons—the means to wage war—it is China's newfound status as a significant and growing net oil importer that may have the greatest impact on security and stability in the Pacific. Competition over disputed territories believed to possess oil and/or gas, it is argued, provides a potential source of conflict. China's energy choices will determine whether oil becomes a source of tension or a spur to China's deeper integration into the global economy. China has gradually moved toward markets to deal with its energy problems. Its capacity to meet current and looming energy challenges of the magnitude discussed here will almost certainly depend on the degree to which it sheds much of its remaining statist distortions and completes the journey from Marx to market.

China is already the world's second largest consumer of energy, though its per capita energy consumption is still minuscule compared to that of the United States (less than one-tenth) or South Korea (one-third) and barely half the world average. It is difficult to overstate the significance of such facts. At this "'take-off'" stage of development, energy-intensive industries (chemical, petrochemical, iron and, steel, heavy construction) play a prominent role. This is true already, though oil and gas together account for only 22 percent of China's energy consumption. Moreover, China's per capita oil consumption is only 8 percent of Japan's. If China's per capita private automobile ownership quintuples by 2015 (which would still be less than four autos per hundred people versus 75 per hundred in the United States) the increase in world oil demand would be substantial.[1] Using a conservative growth figure, the moderate-growth IEA scenario of 3.6 percent energy demand growth to 2020, China's economic growth trajectory indicates that it may account for nearly half the growth in Asian demand in the coming two decades. The inescapable conclusion is that China has only recently begun to be a significant factor in world energy markets.

This was in a sense announced in dramatic fashion in 1997, when with little warning, the Chinese National Petroleum Company (CNPC), hardly a name-brand global energy player, stunned the world oil industry by outbidding major Western firms (including Exxon, Texaco, and Amoco) to acquire oil fields and to build large pipelines in Kazakhstan and develop oil fields in Iraq, Iran, Venezuela, and Sudan worth over $8 billion. In the space of a few months, between May and October 1997, it seemed to many, China had come out of nowhere to suddenly demonstrate its ambition and ability to leave a large footprint on the world oil business.

For the first time in its 5,000-year-old civilization, the long energy-insulated Middle Kingdom, with its considerable natural resources and famous penchant for self-reliance, was acknowledging—in bold neon lights—that it required resources from abroad to sustain its daily life. Suddenly, a fledgling state oil company was establishing itself as a global "resource warrior," even though Western firms believe it vastly overpaid for its invest-

Table 5.1 Growth Rates of Real GDP

	2000	2010	2020	2050
Demand	200	60	320	520
Domestic Supply	155	65	180	80
Deficit	45	95	140	440
Deficit met by:				
Substitution Fuels	—	10	51	280
Oil Imports				
Projections 1996*	45	85	89	160
Projections 1999**	35	60–75	130	—

*PRC (1996) *China Energy Strategy Study (2000–2050)*, Beijing, (in Chinese).
**China Oil, Gas and Petrochemicals Newsletter,* vol. 7, no. 24, December 15, 1999, p. 1
Source: "China's Worldwide Quest for Energy," OECD/IEA Publication, 2000.

ments.² It left many in both the oil industry and the field of international security with a discomfiting sense that something new and unanticipated was astir. Did China have some new grand strategy melding energy and security?³

This development, actually one with myriad meanings, is a useful window into the complex question of energy in China, an issue where, like much of what goes on there, things are rarely what they appear to be. The CNPC large-scale overseas oil investments (discussed further below) were indeed a signal that Beijing had abandoned the hope that its own considerable oil resources would prove to have enough capacity to fulfill its growing needs. Moreover, the assertive purchases also signal the rapid growth of—and increasing importance of—oil in China's energy mix. (See Figure 5.1.) They also illuminate Beijing's sense of urgency in bolstering its energy security by reducing its vulnerabilities to a growing oil deficit. The CNPC ventures may also have shed light on a new importance attached to gaining a Central Asian foothold as a means to develop energy resources for China's interior, part of China's grand notion of its vast territory as an energy "land bridge."⁴ In this case, it appears that CNPC's overseas investments reflected congruent interests of Beijing's economic and national security bureaucracies. It is also another indication that state-run firms are not yet completely autonomous actors but also reflect foreign policy interests.⁵ Yet even if the CNPC moves were part of a larger strategic energy agenda, the oil investments also reveal China's goal of—and the enormous obstacles to—transforming fledgling state energy companies into competitive multinational firms (discussed below). For example, three years later, the Kazakhstan pipeline deal was postponed indefinitely. In addition, it seems that China's refineries do not yet have the capacity to process much of the low quality crude from Persian Gulf fields of Iran and Iraq in which CNPC invested.⁶

For a variety of reasons (see below), CNPC may have overreached. By the end of 1999 speculation in Beijing was that as the problematic realities of capitalizing on the resource acquisitions became more apparent, Prime Minister Zhu Rongji was troubled—and taking names—over billions of dollars committed to projects with questionable economic cost-effectiveness.⁷

The Context of Reform

Nonetheless, the acquisition of equity oil stakes abroad in the 1990s as China became a net oil importer were, in retrospect, an opening gambit in China's quest for long-term energy security. It is one piece of the puzzle

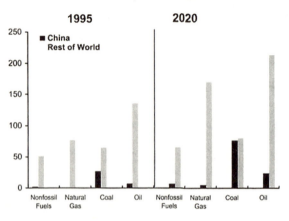

U.S. Energy Information Administration, "International Energy Outlook, 1998."

Figure 5.1 Total Energy Consumption by Fuel: China and the Rest of the World, 1995 and 2020
Source: U.S. Energy Information Administration, "International Energy Outlook, 1998."

that is China's energy policy—or more precisely, its search for a coherent, viable energy policy. The aggregate numbers of China's considerable resources and its enormous needs (outlined in the previous chapter and discussed below) are imprecise, and in any case, only begin to tell the story. The energy sector cannot be fully understood outside the context of: economic reform in China; the domestic geography of energy; indeed, of an entire society in an unprecedented state of transition as China in its helter-skelter fashion gradually integrates itself into the global economy in its quest to modernize.

Though it has some idiosyncratic features, the energy sector broadly mirrors the improvisational quality of China's economic reforms, as previously noted in Deng Xiao-ping's colorful imagery, "crossing the river by feeling for stones." Put less gracefully, it tends to move two steps forward, one step back. But further market-oriented change in the workings of the energy sector will be critical to China's ability to meet its energy needs. The decidedly nonlinear process of restructuring the state-owned energy firms is particularly complicated by the strategic importance attached to energy—in terms of both national security, and impact on the economy—and the magnitude of the challenge. China's economic success over the past two decades—growing by an average of 9.5 percent annually from 1980 to 1995—generated a widespread euphoria, suggesting to many that China was yet another example of a widening Asian economic miracle.

Such simple optimism increasingly appears unwarranted. Many projections of China's energy demand over the first two decades of the twenty-first century are premised on growth rates of 8 percent or higher. Yet China begins the new century with the fate of its economic reforms highly problematic, with 6 percent growth or lower quite probable for the near term.[8] In fact, China's growth in the 1980–96 period may say as much about the stultifying impact of Maoist development policies on the economy as it does about Beijing's ability to achieve the unprecedented feat of transcending the legacy of collectivist economics, the "iron rice bowl," and become a modern, market-based, industrialized society.

Initial reforms, allowing private enterprise, particularly in rural areas and special economic zones in coastal areas such as Shenzhen and Zuhai, along with opening up to foreign investment after 1978, unleashed entrepreneurial energies that catapulted China's economy forward. But after two decades of reform, in 1998 some 87,000 state-owned enterprises (SOEs) employed nearly two-thirds of urban workers, absorbed over two-thirds of investment capital, though they produced some 30 percent of national wealth, down from 70 percent in 1978. Moreover, roughly half the SOEs, were officially listed as loss-making, up from 26 percent in 1992.[9] This figure is misleading, and may be understated because of government-directed lending by China's state banks: many SOEs, while nominally reporting profits, actually destroy rather than create value. As Edward Steinfeld has argued, "Because of opaque accounting practices and implicit subsidization through working capital loans, insolvent firms have been able to continue production."[10]

During the first two decades of reform, China's banks nearly doubled credit, from 53 percent of GDP in 1978 to 100 percent of GDP by 1997.[11] China has used its four major state banks to finance SOEs with politically directed lending. All the banks are, despite initial recapitalization efforts begun in 1998, believed to be insolvent. This may have in effect created something resembling a giant Ponzi scheme, funneling money to sustain production, generate illusory statistics, and maintain employment.[12] In late 1999, China's audit office revealed that two of its four largest banks had overstated their assets by a staggering $48 billion, with implications of widespread corruption.[13] By 1995, the debt-to-equity ratio of many state-owned firms was 500 percent, a level rivaling that of the worst Korean and Indonesian firms affected by the 1997 financial contagion. This problem has remained a manageable, largely domestic affair, as China's currency is not convertible, and few loans were from foreign lending institutions. But such financial flows have led to massive overcapacity, which helps explain continued deflationary pressures for 28 months by the end of 1999. It has also led to a buildup of bad debt, estimated at between 25 and 40 percent of China's GDP, though it is unclear whether even the financial institutions themselves know the exact figures.[14]

This state of affairs created a sense of urgency in Beijing, with an implicit understanding that it was not possible to remain, in effect, a little bit pregnant: to sustain Chinese economic growth, more market dynamism was necessary. Reforms needed to move forward, in the energy sector as well as the economy writ large. At two important decision points, a September 1997 Communist Party Plenum, and the March 1998 Ninth National People's Congress, Premier Zhu Rongji launched an ambitious agenda for a second wave of reforms. Through privatization, corporate consolidation, mergers and recapitalization, and reorganization in the banking sector, Zhu has sought to accelerate market-oriented changes of medium and large SOEs, and to shield China from the potential effects of the type seen in the Asian financial crisis by closing down loss-making SOEs and weak banks.[15] In the energy sector, this meant a reorganization of the major state-owned firms as well as price reform (see below) as part of a broader program to accelerate reform.

China's overall reform push involved a broad effort to rationalize and consolidate economic and financial matters by—at least temporarily—regaining more centralized control over commercial and financial dealings that had devolved to provinces and localities. The effort also targeted local corruption, which had become rampant and pervasive. In 1999, for example, when Beijing cracked down on regional investment trust companies, known as the ITICs, the largest, Guangdong's ITIC, had only $785 million in assets but $4.7 billion in liabilities.[16] Similarly, Beijing fired 1,179 local prosecutors for corruption in 1999, as well as revealing that $14 billion, more than 20 percent of its national revenues for 1998, was unaccounted for.[17] This may be anecdotal, but they are telling examples of the extent of rot in China's political economy, particularly at the local and provincial level—and of Beijing's efforts to clean it up. Another aspect of, and to some degree, a byproduct of, these reform efforts is ameliorating massive environmental problems—China has the dubious distinction of having seven of the ten most polluted cities in the world, according to a 1998 World

Health Organization report, with acid rain falling on 30 percent of China's total land area.[18]

These realities reflect Beijing's messy dilemma of how to dismantle the legacy of a half century of communist economics while maintaining social and political stability—that is to say, the legitimacy of the Communist Party political monopoly. As China's accession to the World Trade Organization underscores, Beijing knows that the next wave of reforms needed to sustain economic growth will require dismantling what oxymoronic official ideology calls a "socialist market economy." This is true for a wide range of industries, particularly the oil industry. China's WTO accession will mean a commitment to phase in reforms involving many activities of state-owned monopolies such as CNPC and Sinopec. This will involve tariffs, pricing, quotas for some oil products, and the opening of the Chinese energy market to foreign ownership, distribution, and retailing (e.g., gas stations). While this may mean some short-term pain for Chinese monopolies, the result will be a more competitive energy industry. This is a microcosm of the rationale for the WTO agreement Beijing signed, with the intention of the WTO commitments becoming an external source of pressure to accelerate what are, under any circumstances, very difficult reforms.

To fix the financial system it must reform and privatize SOEs. But to privatize SOEs it must also accelerate efforts to create an alternative social safety net, since in China, state-owned enterprises, specifically, their *danwei* (work units), are responsible for administering social welfare—housing, education, health care, pensions. And China must do all three roughly simultaneously. Otherwise the mutual and interconnected burdens of each of these three elements would tend to make market-oriented reform exceedingly difficult, if not impossible. Multiply that agenda by 1.2 billion people, the vastness of China, and one begins to grasp the enormity of the challenge—and perhaps help explain why implementation has been slow. While few expect these goals to be achieved in anywhere close to the three years that Beijing initially proclaimed, there has been a demonstrable effort to streamline the PRC government apparatus and reform the SOEs (which by some estimates), led to as much as 20 percent of SOE workforce downsized in urban areas by the end of 1999. Nonetheless, this agenda has proven enormously complex and difficult to achieve without exacerbating social unrest. One result, is, for example, that Beijing has indulged in large-scale deficit financing to minimize the dislocation of its reform agenda, more than $50 billion in 1999.[19]

Energy and Reform

In broad terms, China's reforms have meant significant decentralization of authority, with provincial and local governments assuming more financial responsibility for unemployment, education, health care, pensions and other social services. This has generally meant more fiscal responsibility and autonomy for local authorities, as well as lower tax receipts sent to Beijing. Not infrequently, it has meant competition between different levels of government. Indeed, it can be argued that China has in an ad hoc fashion developed a system that resembles a kind of informal—if still somewhat ad hoc—federalism. The reform process has been characterized as having "become too complex for authorities in Beijing to control." According to this logic, three trends lock in the trajectory of reform: "the construction of dense networks of contractual and ownership ties between Chinese enterprises to internalize both market costs and political risks; a withering away of the central government's capacity to monitor the gradual privatization of Chinese companies; and the rising economic authority of fiscally autonomous local governments that compete with each other to provide market environments suitable for investment."[20]

In the energy sector, broadbrush policy direction (e.g., degree of foreign involvement) and very large national investments such as major national infrastructure (e.g., pipeline) generally emanate from Beijing as does final approval on other key decisions, with the People's Liberation Army and Politburo weighing in with national security considerations. But implementation of specific energy projects are the result of an interactive process, or less charitably, of case-by-case, closed-door bargaining process between various levels of authority. This process frequently is pushed from the local and provincial levels up, seeking approval at the national level, with numerous actors in the process including local firms, subsidiaries of larger SOEs, state energy companies, in some cases competing with other state energy companies. For example, local subsidiaries may oppose CNPC downsizing decisions for their region. Not infrequently, decisions made in localities or provinces distant from Beijing are made that are contrary to existing policies, or create de facto new policies that are only later formalized by the political hierarchy. The complexity of managing this process is illustrated in one hypothetical example of the workings of the oil and gas sector:

> Oil may be pumped out of the ground by a collective production team under contract to the central-government-run state-owned enterprise that

manages or "owns" the oil field, and from there, it might be shipped via a provincial-government-run, shareholding pipeline enterprise to a traditional municipal-government-owned oil depot, It may then be loaded onto trucks and moved under contract by a nearby rural township and village enterprise to a joint -venture refinery set up by central and local government authorities as a cooperative technological venture with a foreign company. The refined petroleum might then be transported via an urban collective transportation team to a district-government-owned stock cooperative gas station.[21]

It is such intricate inter-actions that comprise the management of the production, processing, and distribution of China's energy resources. And it is in the crucible of these fluid, dynamic economic and social circumstances that China's decision making on how to manage its energy challenges is playing out. The parameters of China's energy choices are also defined by the underlying realities of its relatively new status as an increasingly large-scale oil importer, its still burgeoning demand for electricity, and its growing environmental pressures.

At the outset, the course China pursued on energy has its constraints and challenges fundamentally shaped by geography and geology. Coal has accounted for roughly 75 percent of China's primary energy mix for the last quarter century (down from 93.9 percent in 1960), though dropping sharply to 68 percent in 1999 (due to overcapacity, lower growth and shutdowns of a host of mines). Oil accounts for about 25 percent, hydro (2.5 percent), natural gas (2.5 percent) and nuclear (1 percent).[22] China's natural endowments feature large quantities of the environmentally wrong type of resources (massive coal reserves) in the wrong places (the North and West) and the strategically right type of resources (oil) in the wrong places (North and West), while China's largest and fastest-growing energy demand is in the southern and eastern coastal provinces. (See Figure 5.2.)

This geography has reinforced trends of decentralization in China's economy and shapes China's energy options. Whether it is gas-rich Szechuan, oil-rich Heilongjiang, coal-rich Shanxi, or energy-resource-poor Guangdong, each province has its own unique energy imperatives, its own economic logic and sense of self-aggrandizement, and its own set of ideas of how to utilize and/or obtain resources to meet its current and anticipated needs. China's geography/geology also tends to reinforce this complex and oft-misunderstood process of decision making, one with particular characteristics in the realm of energy, and forms a patchwork quilt that comprises China's approach to energy.

94   *The Asian Energy Factor*

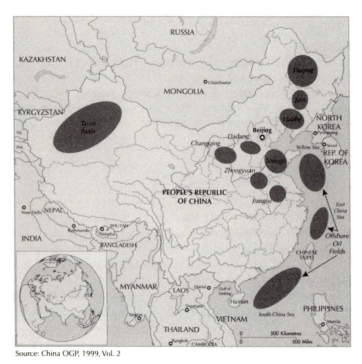

Source: China OGP, 1999, Vol. 2

Figure 5.2 Map of China Indicating Major Provinces and Oil Fields
Source: "China's Worldwide Quest for Energy," OECD/IEA Publication, 2000.

The Challenge of Supply and Demand: Mixed Blessing of Coal

While it is China's burgeoning oil needs and their geo-strategic implications that capture global attention and intersect its foreign policy, the story of energy in China has largely been one of coal. The coal industry exemplifies the impact of economic reform on China's energy situation and underlines the importance of the environment in the calculus of energy choices. For more than a decade, since it surpassed the Soviet Union in 1989, China has been the world's largest producer (one-third of the world total) and consumer of coal, and has the world's third largest proven reserves, after the United States and Russia, some 115 billion tonnes. China possesses just under 12 percent of the world's total proven reserves, enough for about 95 years at current rates of consumption.[23] Some estimates of total reserves range as high as 300 billion tonnes, which would suffice for 225 years at current rates of consumption.[24] Much of China's coal, particularly in the North and Inner Mongolia is low-sulphur, while lower-quality coal with more limited uses is concentrated in central and western China.

These abundant coal resources comprise 72 percent of China's primary energy mix. In contrast, coal comprises 25 percent of U.S. primary energy use and 20 percent of Japan's. China's economy remains heavily dependent on coal. Li Peng, former premier and chair of the National People's Congress (NPC) conceded in a major article on energy policy that, "Not much change in this situations is expected for a long time to come."[25] Despite substantial environmental pressures, the most optimistic outside analysts' projections do not expect coal's share of China's energy mix to fall below 60 percent by 2020, though this is Beijing's stated goal.[26] Over 60 percent of China's primary energy is for industrial use, with nearly half of coal used in industrial boilers and furnaces; coking for steel production is also a major use, and until the late 1990s, when major cities slowly began to shift to gas, coal, and more recently, coal briquettes, has tended to be used for residential cooking and heating.

Coal comprises nearly 40 percent of the goods transported on China's overburdened and dilapidated railroad and inadequate road networks. Very little of China's coal is cleaned or even sized, further exacerbating massive pollution problems. Moreover, fatal mining accidents are some 30 times more frequent in China than the OECD average.[27] Some 21 million tons of sulphur dioxide are emitted into the atmosphere annually along with 14 million tons of smoke-filled dust and another 13 million tons of particulates.[28] Clean coal technology is rarely employed, and less than 20 percent of coal is even washed, largely for economic reasons. One hopeful development that may accelerate modernization is increased openness to foreign investment in the coal industry. Techniques such as conversion to gas and liquid fuels suggest that China may be moving, if gradually, toward reducing the tension between its coal dependence and environmental concerns. One of the most hopeful signs is investment by ARCO and Texaco in coal-bed methane production (which reduces pollution). U.S. firms believe there is great potential to develop coal-bed methane, most concentrated in northern China, with the largest impediments being lack of infrastructure, and Beijing has set up a coal-bed methane firm as well.[29]

China's coal industry has been languishing in oversupplied markets in recent years, with production dropping from 1.4 billion metric tons in 1997 to an estimated 1.1 billion in 1999. Large state-owned coal mines, rank among China's loss-making SOEs. More than a dozen large mines filed for bankruptcy in 1999, and a number of mines have been closed. Since the mid-1980s, nearly half of China's coal has been produced at some 75,000 small, inefficient (many illegal) mines owned by localities, collectives, or individuals. This situation may be in part the result of Beijing's decontrol of prices, partially in 1984, and fully in 1996. This created a rush for quick profits by small mines, using primitive extraction techniques. However, in 1998, Beijing began a crackdown, designating 25,800 small mines for closing.[30] Yet despite this more market-oriented pricing, transport costs, which are not transparent still leave the coal industry with competitive problems.

As coal demand is projected to rise significantly by 2020, there are efforts to develop coal-fired power plants located around large mining sites. China is also obtaining new turbine power technology, which could facilitate large (600 megawatt) power plants, significantly increase energy efficiency, and reduce coal dependence. Nonetheless, the number of thermal 300 to 600 MW plants completed has doubled since the late 1980s. A process of decentralization and deregulation of the electrical power industry has been under way since the 1980s, accelerating since 1996, when a legal framework for the power industry was adopted. China has benefited from substantial foreign direct investment—nearly $18 billion—into electricity production (thermal and hydro) and more than a quarter of World Bank loans to China have been energy-related.[31] Yet there remains price distortion in its electricity market, with rural areas paying much more than large urban areas, for example. One interesting question is the degree to which China will follow the global trend of deregulation of power supply and allow independent power providers (IPPs), particularly in joint ventures with foreign investors, free reign and a competitive, level playing field with state-owned power plants.

Emerging Power Sources: Hydro and Nuclear

As discussed in the previous chapter, the rapid expansion of China's middle class has meant an explosion of electrical appliances—refrigerators, washing machines, televisions, fans—in urban households since the mid-1980s. Moreover only about 80 percent of China's populace is hooked into an electrical grid. One major study of China's electrical power options forecast up to a fourfold increase in power demand between 1995 and 2020.[32] As China has sought to keep pace with rapidly expanding demand for electricity, which at its current stage of industrial development tends to grow faster than other forms of energy, it has only gradually begun to look beyond hydrocarbons. Coal still remains the source of more than 70 percent of China's electricity generation. Renewables such as wind power and solar energy are negligible energy sources, playing a niche role for isolated locales.

Nonhydrocarbon resources, particularly hydropower and nuclear energy, do play a significant and growing role in China's electricity expansion plans. China's vast river network endows it with the world's largest potential exploitable hydro resources, nearly 380 billion watts. However, hydropower only accounts for 5 percent of China's energy use at present. As of 1998, China had only about 60 gigawatts of installed hydroelectric capacity. The Asian economic crisis contributed to a slowing of the Chinese economy in 1998 and early 1999, creating short-term oversupply. In response, Beijing sought to close down small thermal plants (along with small coal mines as noted above) and slowed approval of new power plant construction, including a freeze on foreign built generators of less than 600 MW capacity to protect China's domestic manufacturers.

Large fixed investment costs and the expense of transmission lines to population centers required by hydropower make thermal energy appear a less expensive near-term alternative. But Beijing has decided to pursue an assertive program of large-scale hydroelectric projects, with the State Power Corporation (SPC) announcing in September 1999 it would invest $7.23 billion in five major projects by 2010 designed to add a total of about 17 gigawatts of electrical power in.[33] This is in addition to 19 already operating hydroelectric projects generating more than one gigawatt, and power projects under construction in 1999 that are to have a planned operating capacity of 32 gigawatts. To put this in perspective, China's electric power capacity is expected to grow by 300 gigawatts between 1992 and 2010. This will require substantial investment capital, as the rule of thumb is that each gigawatt requires about $1 billion in investment.[34]

The magnitude of projected demand may help explain the controversial giant Three Gorges Dam project now under construction. The idea reaches back almost a century to Sun Yat-sen and the dream of curbing the devastating flooding of the Yangtze River valley—which took more than 3,000 lives and destroyed 5 million homes, causing $36 billion in damage in 1998. Three Gorges is a particular favorite of some Soviet-trained engineers in China's top echelon of leadership, such as former premier and current head of the National People's Congress Li Peng. It is the world's largest hydroelectric project, designed to generate 18.2 gigawatts of power with 26 separate 700 megawatt generators. Scheduled for completion, in 2009, at a cost estimated officially at $25 billion, the project will be 607 feet tall and span 1.4 miles across the Yangtze, creating a 370–mile-long reservoir. However, the timetable for completion appears overly optimistic and the official cost estimate, as is frequently the case worldwide with megaprojects, appears significantly understated.[35]

The project has sparked controversy both inside and outside China on several levels. Environmentalists fear that the huge dam will dramatically increase water pollution along the Yangtze, some argue it may clog the major river channels of the huge central Chinese city of Chongqing, and others are concerned that it will flood the habitats of endangered species. Another layer of controversy revolves around the need to forcibly move and resettle about 1.2 million people and some 1,300 archeological sites. Despite these controversies, European and Canadian export-import banks have backed their respective home companies efforts to bid on construction contracts for Three Gorges, though the U.S. EXIM bank has refused to do so. The three-phase project is now in its second phase, which is due for completion in 2003. The controversial nature of the project of such enormous magnitude in a society in flux suggests that the successful completion of Three Gorges as planned is hardly assured.

Nevertheless, Three Gorges is a symbol of both Beijing's long-term effort to develop more of its hydroelectric potential and its determination to keep pace with power demand. China, however, is still a long way from fashioning a competitive electric power market, which it seeks to do over the longer term. China ended the twentieth century with foreign investment permitted in power generation but not yet power transmission. As noted above, it is unclear how and when Beijing will adapt the sort of sweeping deregulation that is becoming a global trend in the power industry, led by Germany and the United States. One element of such a power market will be facilitated if China achieves its goal of creating a national power grid to which plants could sell power at market-based prices. Again, energy policy and the fate of economic reform appear closely intertwined.

Limited Nuclear Future
There has been far less momentum in the development of nuclear power in China. Despite nuclear's attraction as a clean fuel and China's ambitious expansion plans, and Western hopes of $100 billion markets, nuclear power appears to be expanding only incrementally. Nuclear power currently provides just over 1 percent of China's electricity. China's nuclear path is unlike that of many current and wannabe nuclear proliferators. Most have used the cover of civil nuclear power; some, such as Iraq and North Korea, joined the Non-Proliferation Treaty (NPT), obtaining nuclear know-how and fissile material with which to pursue nuclear weapons. For China, developing functioning civil nuclear power was not a stepping-stone to

attaining nuclear weapons. Rather, it was not until the 1980s, nearly two decades *after* China developed nuclear weapons in 1964, that it first looked to nuclear power to boost its electricity-generating capacity. In China's energy strategy, nuclear power was conceived as part of a policy response to the disproportionately large and burgeoning power needs of resource-poor coastal and eastern provinces. But China's two nuclear power plants did not begin commercial production until 1993. The largest impediments to expanded nuclear power use are currently the long lead time (roughly 5–7 years) and the high startup costs of nuclear power—cost estimates range from a high end of roughly three times as much as coal per kilowatt to construct a nuclear plant to 70 percent less than coal using scrubbers.

China's first nuclear power plant, a 300 MW unit at Qinshan in Zhejiang province started commercial power production in May 1993. The first of two larger, 900 MW power reactors at Daya Bay near Hong Kong in Guangdong province came on line in August 1993, with the second beginning operation in May 1994. Two more reactors are under construction at Qinshan, with two 700 MW Candu heavy water reactors planned for a third phase. A third nuclear power plant with two large French-built 1,000 MW reactors at Lingao near Daya Bay is scheduled for construction by 2001, and a fourth plant utilizing Russian reactors is planned for Lianuyngang in Jiangsu province. The additional plants under construction would nearly double China's current operating capacity of 2.1 gigawatts if the projects are completed and go on line by 2003 as planned. Guangdong, whose economic dynamism accounts for over one-third of China's exports, views nuclear power as a key component of its energy strategy, with officials hoping that nuclear power will supply some 20 percent of the province's electricity by 2025–2030.[36] Officially, China's longer-term goal is to boost its nuclear capacity tenfold, to 20 GW by 2010 and 40 GW by 2020. Yet as of mid-2000, increasing fiscal constraints appear to have put these nuclear power ambitions on hold, as no additional funds have been allocated for construction of such facilities. Both CNNC and Guangdong provincial officials indicate that completion of these plants and more broadly long-term national plans are problematic.[37]

However, Chinese officials still see nuclear power as an important component of their long-term strategy to reduce dependence on coal to 50 percent or below on the one hand, and imported oil or gas on the other.[38] Chinese scientists at Qinghua University's Institute of Nuclear Energy[39] are trying to build China's first high-temperature gas-cooled reactor (HTGR), a state-of-the-art nuclear reactor that is more efficient, much less expensive, and can accelerate the pace of building a nuclear power station in two years. This has been a longtime quest in the nuclear industry, though even if, as advertised, China builds a laboratory prototype for a next generation of nuclear power, it remains to be seen whether it can translate it into an operable, full-scale reactor.[40]

Apart from financial constraints, in the future China is likely to face many of the obstacles that have constrained the use of nuclear power elsewhere. There is the problem of long-term storage of nuclear waste. Daya Bay reactors produce 20–30 tons a year, according to local officials.[41] This raises the issue of public acceptance, which has frozen U.S. nuclear power, is limiting expansion plans in Japan, and has been the object of political contention in heavily nuclear-dependent Taiwan (33 percent of its electricity) as well. Already, there have been protests against the building of a third nuclear power plant in Guangdong by environmental activists.[42] As the 1979 Three Mile Island event in the United States, or more recently, the 1999 Tokaimura accident in Japan, one nuclear accident can have grave implications for nuclear power plans and operations. As evidenced by the proliferation of environmental nongovernmental organizations in China, public pressure is increasingly likely to be a factor in decision making on nuclear power in any case. Against this backdrop, IEA estimates of 11 GW nuclear capacity in 2010 and 20GW by 2020 appear overly optimistic. Looking out to 2010, it is unlikely that nuclear power will supply more than 1.5–1.8 percent of China's primary energy, and by 2020, no more than about 3 percent of China's energy consumption.

Over the coming two decades how China develops its mix of coal, hydro, nuclear, and, increasingly, natural gas will, of course, be critical to China's ability to meet future demand for commercial and residential power—and to addressing its grave environmental problems. Those sectors will continue to comprise a large part, perhaps 75 percent or more of total energy demand, which is likely to double to the equivalent of roughly 40 million barrels of oil per day by 2020. As Figure 5.3 shows, reducing dependence on coal to below 42 percent of China's primary energy, something unlikely before 2050 (barring the introduction of radically new technologies)—and even then, a highly optimistic goal—remains a major challenge to Chinese energy planners.

Part II: New Oil Realities, Gas Hopes

Competing with—and quite likely surpassing—electric power demand as the fastest growing part of China's energy sector over the next two decades will be transport. As tends to be the case in rapidly developing and urban-

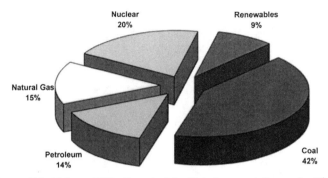

Figure 5.3 China—2050: Sustainable Development Scenario Electricity Generation Primary Fuel Mix
Source: William F. Martin, "Global Energy Security, China, and the Role of Nuclear Power," Prepared for the Senate Committee on Energy and Natural Resources Oversight Hearing on Peaceful Nuclear Cooperation with China, 1997.

izing economies, the demand for mobility skyrockets. EIA projections are that nearly half of the growth in world energy demand over the next decade will be in the transport sector. In contrast to OECD levels (33 percent) and other developing countries (23 percent), the percentage of total energy consumption used for transport by China is so low (9 percent) that the IEA argues that the data is underestimated.[43] Nonetheless, even if the real figure were double the size of China's official data, it would still be relatively small. This peculiarity is in some measure the result of the legacy of central planning in China, which saw Beijing invest in heavy industry rather than light industry and infrastructure. Regardless, the surge in transport needs translates overwhelmingly into a swelling demand for oil, which in turn accentuates the relatively novel Chinese concern about energy security. (See Figure 5.4.)

Concern about its oil balance has loomed ever larger in Beijing's energy security calculus since China became a net oil importer in 1993, even as it was the world's fifth largest oil producer. Yet the production of oil is a relatively new phenomenon in China: it only began to be a significant oil producer in the 1960s, after its Daqing field—still the largest of producing Chinese oil fields—was discovered in the northern reaches of Heilongjiang province in 1959. The aging Daqing field, once a showcase of Maoist "self-reliance," still produces about 1.1 million b/pd, roughly one-third of China's total production. Together with two other mature fields, Shengli and Liaohe, these three fields account for nearly two-thirds of Chinese production. Apart from secondary fields in Xinjiang's Tarim Basin and offshore

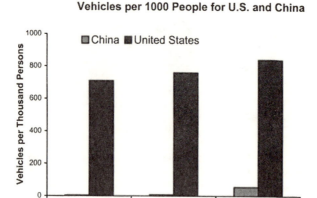

Source: J. Dargay and D. Gately, Income's Effect on Car and Vehicle Ownership Worldwide/ U.S. Energy Information Administration, "Country Energy Profile: China," October 1997.

Figure 5.4 Vehicles per 1000 People for U.S. and China
Source: U.S. Energy Information Administration, "Country Energy Profile: China," 1997.

fields, the remainder of Chinese production is scattered among more than a dozen fields producing less than 100,000 b/pd, most in the north or northeast.

Chinese domestic oil production grew modestly, (averaging just under 2 percent annually) during the 1990s, reaching 3.2 million b/pd in 1997.[44] That made it the world's fifth largest oil producer. But it was still some 600,000 b/pd short of meeting China's demand, which grew that year to some 3.8 b/pd. Moreover, few analysts expect more than modest expansion of Chinese domestic production over the first two decades of the twenty-first century. China's reserves-to-production ratio is 20.5 years, half the world average. A mean projection derived from forecasts of five prominent oil consulting firms and the U.S. Department of Energy suggests 3.6 million b/pd by 2010; the most optimistic forecast is 4.1 million b/pd.[45] Industry analysts and Western oil companies believe there is little prospect of expanding production of the three large fields that account for nearly two-thirds of Chinese production. Indeed, it may require significant investment to sustain production at close to current levels over the coming two decades.[46]

The most likely sources of domestic expansion of China's oil supply are the Tarim Basin in Xinjiang and offshore, particularly Bohai Bay and in the South China Sea. It should be noted, however, that beginning in the 1980s, China's state-controlled oil industry began to open selectively to foreign investment. Western companies have spent more than $2.5 billion since 1982 on oil exploration in China, particularly offshore in the East China Sea and the South China Sea as well as in the Tarim Basin in Xinjiang province.[47] Chinese officials have had high hopes that the Tarim Basin would become their version of Kuwait, with estimates as high as 150 billion barrels of oil.[48] However, the highest-end estimate of potential reserves in Tarim cited by the IEA is 80 billion barrels.[49] A number of exploration blocks in Tarim have been opened to Western and Japanese investment, but thus far, it has only produced about 80,000 b/pd; indeed, all of Xinjiang province has yielded just under 300,000 b/pd, about 9 percent of Chinese production.

While Tarim may have large potential reserves, what oil has been located to date is in isolated, hard to extract deposits. Tarim is in a remote, barren desert environment, with oil workers fighting off biting winds blowing tall sand dunes. Moreover the lack of infrastructure requires investment in roads and telecommunications to develop any oil resources, and nearly 4,000 kilometers worth of pipelines to get it to markets. Yet in mid-2000, Beijing appeared to be renewing its efforts to realize the 4,200 km west-east Tarim Basin pipeline, which would deliver oil to southwestern and eastern China, and reviving the "Silk Road" oil pipeline idea. It is unclear whether this is the result of new developments that promise larger amounts of oil that are more easily obtainable, or a political impetus stemming from its campaign to put more emphasis on developing the Western and inland parts of China.

Similarly, despite extensive exploration by Western firms in the East and South China Seas, only small pockets of oil have been found, many not commercially exploitable. One focus of current exploration activity is in the Bohai Sea area off the coast of Tianjin, which may have more than 1.5 billion barrels of reserves.[50] While modest new offshore finds are probable in the future, they will have little impact on Chinese oil balance.[51] It is frequently reported (and widely assumed) that territorial disputes between China and six ASEAN states over the Spratly Islands in the South China Sea, as well as the Sino-Japanese dispute over the uninhabited Senkaku/Diaoyutai Islands in the East China Sea, revolve largely around massive oil and gas resources. These issues are discussed further in chapter 9. But the reality is that while the well-explored area surrounding the Spratlys appears to be gas-prone, no major oil company believes there are more than modest reserves of commercially exploitable oil or reserves (less than 200,000 bp/d). Similarly, there is no evidence of substantial oil reserves in the 200 mile economic zone off territorial waters of the disputed Sino-Japanese islands. In sum, any way it is added up, there appears little prospect for more than marginal increases in Chinese domestic oil production in the near future.

Yet even as China's economy slowed down from the dizzying double-digit growth of the late 1980s and early 1990s, its oil demand continued to swell. China's oil consumption nearly doubled from 2.2 m b/pd in 1990 to an estimated 3.9 to 4.1 m b/pd in 1999.[52] China's current import needs are still relatively modest, in the 600,000–900,000 b/pd range. Given the problematic character of China's economic reforms, projecting long-term growth is obviously a hazardous exercise, particularly over the coming five years. But even using conservative projections of 3.6 percent GDP growth, China's oil needs jump markedly by 2010, to roughly 5.5 m b/pd. GDP growth of 5.5 percent would increase China's total consumption by 2010 to 6.6 m b/pd. By 2020, China's oil demand can be expected to rise to 7 m; 8.8 m b/pd if growth is 5.5 percent. Thus, by the end of the decade China will need to import at least 2 m b/pd, and by 2020, 4–5 m b/pd, or more than the entire production of Kuwait.[53] Moreover, it is conceivable that in the period from 2005–2020 (if its current reform agenda is realized), China's real annual economic growth could rise to the 8 percent range, which would boost its oil demand to roughly 11 million barrels.

Per capita oil use in China is nearly 30 times less than that of the United States, which offers a hint at the magnitude of its potential growth in demand. The data is admittedly anecdotal, but in discussions with a cross section of urban Chinese, what surfaces quickly on the wish list of many for their next big consumer item is an automobile. In 1994, China's state planning bureaucracy unveiled plans to produce a "people's car," a compact auto for which Beijing calculated, not altogether incorrectly, that major multinational auto companies would compete to produce.[54] This was a pet project of then Prime Minister Li Peng. The project is proceeding in piecemeal fashion, and does not yet appear to have grown to the scale of such a national project, though a number of production and assembly arrangements with Western firms (most prominently, Volkswagen in Shanghai) have proceeded. Indeed, at a time of rapid consolidation among global auto giants, there are still more than 100 local auto producers in China, turning out poor-quality products, only five able to produce 100,000 units annually, with different provinces blocking national trade in automobiles.[55]

In any case, passenger auto production has increased incrementally to 350,000 in 1994, and 500,000 annually by 1999, well under the 3 million figure Beijing's state planners envisioned by 2000. From 1978 to 1995, the number of passenger vehicles per thousand people increased nearly 1,000 percent from 0.3 to 3.2.[56] Exact numbers are difficult to verify, but by 2000, private ownership accounted for no more than about 10 percent of the autos on Chinese roads.[57] Imagining potential markets in China is a historic conceit of the West. Nonetheless, the possibility of up to 250 million cars in China by 2050 if, not the most probable scenario, is not unimaginable. (In 1999 there were some 700 million cars worldwide.) Moreover, even were growth in private autos in China to be only a fraction of that, the impact both on oil demand and the environment would be staggering, considering the astonishing levels of pollution already in major Chinese cities. It would also require massive road construction: already rush hour in major cities rivals that of Seoul or Taipei; and Beijing streets seem traffic-laden much of the day.

To put the mind-boggling projections in perspective, contrast China's private auto ownership of 3.5 per thousand in 1999, to 27 for Thailand. Yet new technologies might temper the impact of large-scale private auto ownership in China both on the environment and in regard to total oil demand. By the time China's consumer auto market begins to take off, it is probable that the a large fraction of autos purchased may be low-emission hybrid gas/electric (probable by 2010–2015) that get 70–80 miles per gallon or fuel cell powered. In any case, growth in the transport sector will by 2025–2030 lead China to overtake the United States as the world largest oil importer.

China's Oil Industry: Managing Exploding Demand

China is looking to CNPC-owned overseas fields to mitigate its sense of vulnerability to foreign oil imports. Over time, China's ability to manage this mushrooming demand (and still faster-growing demand for natural gas) will in large measure depend on the success of structural reform and deregulation of its entirely state-owned petroleum industry. In a series of increasingly market-oriented steps since the mid-1980s, China has sought to transform a centrally planned and tightly controlled industry into several globally competitive state-owned energy companies, perhaps modeled after other government-owned companies like Malaysia's Petronas, or more ambitiously, South Korean *chaebol* conglomerates. But when economic reforms began in 1979, the entire industry was under the control of state ministries. One exception was the China National Chemicals Import and

Export Corporation (Sinochem), which existed to manage the import and export of petroleum products.

Initially, China sought to organize the oil industry along functional lines. In 1982, Beijing created the China National offshore Oil Corporation (CNOOC) to manage offshore exploration and production. China needed foreign investment and access to technology to develop offshore oil, and CNOOC became the vehicle for such international investment, developing foreign partnerships, licensing arrangements, and experience in offshore exploration and drilling as well as absorbing management expertise. Similarly, the China Petrochemical Corporation (Sinopec) was carved out of ministries to control major refineries, the majority in the north and east. In 1989, the Petroleum Ministry was abolished and the China National Petroleum Corporation (CNPC) was formed to control onshore exploration and production. These divisions by operational functions, along with artificial prices, perpetuated market inefficiencies and sparked political contention between the state-owned enterprises who benefited and/or were penalized unevenly by government subsidies.

The Chinese party/government bureaucracy and the state energy companies are still evolving along a path—albeit a bumpy one—of decentralization and deregulation after several phases of bureaucratic and price reform in the 1990s. A two-tier pricing system had required CNPC to sell oil to Sinopec refineries at artificial prices, subsidizing refineries and consumers, making it difficult for CNPC to become profitable, and accentuating market distortions. These distortions led ostensibly specialized Chinese parastatals to begin operating more vertically outside their specialized functions (e.g., challenging Sinochem monopoly on imports) and also abroad in response to, variously, pressure and opportunities from different provinces, particularly the more economically dynamic southern and coastal provinces. Low oil prices in the early and mid-1990s exacerbated these pressures, as the cost of domestic oil transported from north to south (absent domestic transport subsidies) became significantly more expensive than Middle East imports for coastal provinces. As China became a net importer in 1993, its oil business was faced with more complex realities. The petroleum bureaucracy began to respond to the new pressures, on occasions, by slapping quotas on oil imports, which only increased smuggling, which reached alarming proportions.[58] Beijing encouraged an outward orientation for its state companies, and CNPC, CNOOC, and Sinopec began to form joint ventures and overseas subsidiaries in an effort to diversify secure supplies.

In 1993–94, in the first of several phases of restructuring in the energy industry, China's oil market moved from tightly administered control to a more regulatory posture. While Beijing retained the two-tier pricing system, it eased some price controls incrementally. The State Council reorganized the oil industry, removing CNPC, CNOOC, and Sinopec from the Energy Ministry, which was abolished in 1993–94. The state companies were given ministerial status and placed under the SPDC. In March 1998, they were put under the State Economic and Trade Commission (SETC), which Beijing seems to be trying to turn into a Chinese version of Japan's Ministry of Trade and Industry (MITI), providing guidance and direction to economic development. With this structure, it appears Beijing's top political leadership views its power to appoint the presidents of state companies as a key instrument of control. Yet with, for example, CNPC's 51 local affiliates, and its devolving of some operations to 18 Petroleum Administration Bureaus, often tied more closely to local and provincial government, control is clearly a relative term. In 1997, in what appears to have been an effort to spur competition, the Ministry of Geology and Mineral Resources was abolished and renamed the Ministry of Land and Natural Resources; and the China Star National Petroleum Corporation was created out of a number of exploration bureaus and research institutes, though it later merged with Sinopec.

The overseas acquisitions of CNPC, Sinopec, and CNOOC must be viewed in the context of these reforms. As Beijing eased price controls after the 1993–94 reforms, CNPC's revenues shot up dramatically. Indeed, by some accounts CNPC's cash flow tripled between 1993 and 1997, to $21 billion.[59] Flush with cash, and fearful of import dependency, CNPC did what Japan's National Oil Company (JNOC)—with little success—has tried to do since the mid-1970s: obtain its own oil fields to enhance energy security. CNPC bought fields in Canada and also in Peru in 1993, the latter of which delivered its first load of 60,000 tons of crude at the port of Qinhuangdao in Hebei province in 1997. CNPC also agreed to joint ventures with Italy's Agip to develop oil fields in Central Asia and Africa. CNOOC launched its first overseas venture in Indonesia, purchasing a bloc from ARCO, and Sinopec has entered into a number of refinery joint ventures.

It was the explosion of CNPC commercial activity in 1997, however, that grabbed world attention. CNPC paid a premium for fields in Venezuela, Sudan, and southern Iraq (on hold until UN sanctions are lifted), and in Iran signed a joint venture agreement with Iran's National Oil Company to explore offshore oil. It is not unusual for oil companies to acquire assets they may not exploit for a number of years, but that show up in their books as equity. The Iran and Iraq purchases gave China potential sour crude it lacks the refining capacity to process. Chinese refineries can only process about 240,000 b/pd of sour crude, and require large-scale investment in new and upgraded refineries in order to import crude from its new fields.[60] Beijing is aware of this problem and has begun to adapt its refining capacity accordingly, though this is unlikely to be realized before 2002–2003.

Most dramatic, however was CNPC's foray into Central Asia. CNPC committed $4.3 billion to buy a 60 percent stake in Aktybinskmunai Production Association, Kazakhstan's main oil company, and a 60 percent share of Uzen field—all told, resources estimated at about 350 million tonnes. The Kazakhstan investment clearly reflected the intersection of energy security concerns, an initial impulse to "prevent the West from exerting strong influence," as the United States becomes more deeply involved in the region, according to one Chinese analyst. Chinese energy analyst Xiaoje Xu also argues, "China has realized her potential energy interests and geopolitical importance by expanding her geoeconomic space in Central Asia." Xu adds that "China's westward aspiration has been driven by the country's high economic growth combined with increasing energy vulnerability."[61] It was also a perceived investment opportunity. To get the deal, then Premier Li Peng competed with U.S. vice president Al Gore in lobbying the Kazakhs. In addition, to close the deal, China agree to build a pipeline from western Kazakhstan to its Xinjiang province (estimated cost: $4.5 million) and a shorter pipeline from Kazakhstan to Iran.

One impetus for this shopping spree appears to have been a need to spend the cash or have it returned to Beijing's government coffers.[62] Clearly, geopolitics is a factor in China's oil and gas activities in Central Asia. Beijing is very sensitive about the stability of Xinjiang province, where the Moslem and ethnic Turkic Uighur minority has been increasingly restive, and which borders on Kazakhstan, Tajikistan, and Kyrgyzstan. Beijing sees itself as an energy "landbridge," connecting Central Asia and the Russian Far East, though this concept remains to be proven in terms of economic rationality. And some Chinese analysts view Central Asian resources as a means to reduce Middle East oil and dependency. But its prime near-term concern is stability and security of borders. This is underscored in the five-sided 1996 security accord with Russia and three Central Asian republics. Speculation about new "Great Games" is vastly overblown. The Kazakhstan pipeline, which presumably would go from Xinjiang and on to Chinese markets 7,200 km away. This can not be justified on economic grounds. It would make far more sense in commercial terms to export Kazakh oil to Mediterranean markets. A Chinese decision to subsidize such a venture would be a choice based on a security calculus that puts a premium on diversification of supply and/or influence in Kazakhstan, which borders on China's volatile Xinjiang province. As one analysis argues, "The issue for policy makers in China is whether the security benefits of this diversification is worth the cost in terms of higher total import costs."[63] That the Kazakh pipeline decision to move forward was postponed indefinitely, if not outright canceled, suggests that CNPC and China's SPDC are aware of the oil economics.

In early 1998 yet another stage of reforms substantially restructured the industry, part of the wider economic reforms of the March 1998 National People's Congress (NPC) that saw Beijing reduce its total number of ministries from 41 to 29. The most recent reforms reorganized both CNPC and Sinopec into vertically integrated companies. Instead of dividing them on the basis of function, the two state firms were redivided on the basis of geography, corresponding to the location of resources and markets. This involved both SOEs trading assets, with the result that CNPC's upstream oil and gas activities are restricted to northern and western China (roughly 70 percent of onshore production) but it obtained some 30 percent of Sinopec's refining and marketing. Meanwhile, Sinopec gained upstream activities in the southern and Eastern coastal regions (about 25 percent), while maintaining about 65 percent of refining and marketing activity.[64] At the same time, Beijing also unveiled a major price reform initiative, ending the two-tier pricing system. As a Baker Institute study observed, Beijing adopted a pricing system: "linked to the price of foreign crude in the Singapore market. The use of market averages means that domestic prices can vary significantly from world crude oil costs when world prices make larger price swings." But their pricing became more transparent by requiring monthly price statements. CNPC and Sinopec are allowed to alter prices by 5 percent, and as the Baker Institute study points out the new system does not cover all domestic petroleum products, nor are port and transport costs fully factored in. Thus, as the study concludes, "Short-run distortions and inefficiencies are still prevalent, clouding investment decision-making."[65]

These reforms were another important step on the long march to deregulate the energy sector. Beijing sought to reduce the government role in business operations, foster more efficiency, and put its energy SOEs on a more commercial basis. This process will be impacted by China's accession to the World Trade Organization, where energy services are likely to be an issue on the agenda. Nonetheless, it remains unclear how much the latest reforms were merely a reshuffling of the deck. Any way you combine poorly run, uncompetitive, if not oligopolistic, organizations they will only be larger versions of the same problem, particularly when given protected internal markets. By early 2000, for example, CNPC still had 1.5 million employees, and Sinopec, 1.2 million. By comparison, Exxon prior to its merger with Mobil had under 100,000 employees. While CNPC and Sinopec have announced plans to downsize 976,000 employees by 2003, it is unclear when and how this will occur, or whether it will be artificial, in the sense of hiving off subsidiaries to create lower numbers.

It appears Chinese state companies are looking to foreign capital markets to finance their downsizing. This will be a difficult process. CNPC planned to list dual public offerings on the Hong Kong and New York stock exchanges in 2000, followed later in the same year by Sinopec and CNOOC. Yet CNOOC, the most sophisticated and market-oriented of China's energy companies, was forced to delay its offering in October 1999, after dropping the offer price by one-third to $18 a share and still finding no takers. But CNPC hired high-profile Western consultants and hived off profit-making parts of the firm to create PetroChina, which it hopes will induce investors by listing $3 billion in stock, the largest offering to date by a Chinese company on the New York and Hong Kong stock exchanges. Petrochina is 90 percent owned by CNPC, which is stuck with the surplus labor force. In an initial prospectus, PetroChina earmarked the bulk of the IPO for "debt repayment and severance compensation," in effect, saying to investors, "Trust, us, we will become a real oil company."[66] To assuage potential investors, a state council official later denied the reports on how the money would be used, saying that it would be utilized to build the oil and gas business.

Regardless of their success, the point is that the IPOs are another important step in the direction of the market, rather than relying on China's state-owned banks to finance their future operations. Yet CNPC, Sinopec, and to a lesser extent CNOOC are all a long way from commercial viability. The combination of financial pressure from the reforms and new competitive pressures after China's 2000 entry into the WTO may well put to the test their ability to become competitive multinationals over the coming 10 to 20 years. By 2020, China will be in a similar position to that of the United States—dependent on imports for the majority of its oil. Clearly, China sees participation in all aspects of global energy markets as the means to meeting its oil import challenges. Whether China becomes more accustomed to being in such a position, or feels compelled to create a blue water navy to assure the sea-lanes of communication or to enforce territorial claims, is for the near term an open question.

Natural Gas Hopes

Most critical to achieving a more sustainable energy mix—and an important measure of change in its energy sector—is China's ability to develop its increasingly important natural gas resources—and to develop the regulatory environment and infrastructure to absorb gas imports. Beijing hopes to expand gas from 2 percent of Chinese primary energy to 10 percent or more by 2020. Such a fivefold expansion of natural gas would still be far less than half the proportion of gas in the energy mix of either the United States or Europe. Until very recently, environmental concerns were not major factors in Beijing's energy decision, but the horrible state of China's environment—21 million tons of sulfur dioxide, 14 million tons of smoke-filled dust, and 13 million tons of powder-like particularates emitted into the atmosphere each year—have begun to change its policy calculus. The World Bank estimates that the total costs to the central government from pollution is $54 billion, with health and productivity losses at over $20 billion.[67]

In the face of such problems, Beijing sees environmentally cleaner natural gas as an important part of its response, and in 1997–98 made decisions to accelerate the development of its gas industry. But Beijing's moves are only a first step toward a coherent gas policy: changes in pricing, gradually ending subsidies; a developed legal (particularly arbitration) and regulatory structure; allowing competitive bidding for foreign investors; and building pipeline and trunkline infrastructure are a sampling of the impediments to realizing the hoped for expansion of gas in China. On the issue of infrastructure, for example, China has a total of some 6,000 miles of pipeline country-wide: by comparison, the United States has 260,000 miles.

China's gas reserves are believed to be significantly larger than current proven reserves of 1.4 trillion cubic meters (cu. m).[68] Some estimates of proven reserves range as high a 5.3 tril. cu. m. but some estimates of total recoverable gas reserves range as high as 38 trillion cu. m.[69] In addition, China may have up to 35 trillion cu. m. of coal-bed methane. China's gas resources are concentrated in four basins, in Tarim in the far northwest, Qinghai in the west area, nearly 25 percent in Sichuan, the breadbasket in west central China, and Ordos in north central China. There are more modest offshore reserves in Bohai Bay and the East China Sea. It is worth noting that many potential gas-bearing areas remain inadequately explored—even in gas-rich areas such as Sichuan. This has led some analysts to suggest that as much as 100 billion cu. m. a year of gas may be found in China in coming years. China has had some success attracting foreign investment for exploration and production both on- and offshore. As of 1999 over 130 contracts and agreements with roughly $5.5 billion had been signed with 67 companies from 18 nations.[70] China's gas market, however, is still in an embryonic stage, with 19 billion cu. m. consumed in 1998. Until very recently, however, only a small portion of Chinese natural gas was used for power generation or heating; most was used for fertilizer and the chemical industry with prices heavily subsidized. But the combination of economic imperatives and environmental factors have led to major revaluation in Beijing of the utility of natural gas in its energy mix. It was only in 1997 that trunklines enabling gas for cooking in the residential sector were widely employed in Beijing. The government had ordered buses and taxis in Beijing, in other major municipalities, and in Sichuan province to convert to liquified petroleum gas and natural gas by 2000. But to realize its ambitions in regard to gas will require massive investment in transmission and distribution lines and a new set of policies, particularly on pricing, to create an investment climate likely to attract adequate investment.

There are enormous potential markets for gas-fired turbine power generation in China, and some projections suggest that by 2010 more than half of China's gas use will be for power generation.[71] Guangdong province has already begun to expand the use of LNG, and with potential suppliers from Indonesia, Australia, as well as the Persian Gulf, LNG will likely expand significantly over the next two decades. Gas also may figure into emerging energy technologies, particularly gas-to-liquids and fuel cells (see chapter 2). Some leading multinational energy firms envision fuel cells in homes powered by gas as technology that will replace current utilities arrangements over the coming two decades.[72]

Over the past five years or so there have been voluminous studies and articles in trade journals and specialized literature about natural gas in China, about Asian gas grids, and the prospects for Chinese pipelines from Central Asia and Russia.[73] Some of these possibilities for intra-regional energy cooperation are discussed in chapter 9. But few of these schemes are thus far much more than lines on a map. There is no doubt that the use of natural gas and LNG will, over time, expand exponentially in China. But just the process of raising prices to make domestic gas profitable will be a protracted exercise in the 2000–2005 period. Beijing's efforts to build a national gas network connecting the major gas basins will proceed only incrementally. A 503 km pipeline from Sichuan's Zhongxian producing area to Jinzhou in Hebei Province may well be completed by 2002–2003. This will distribute gas to ten cities en route to its final destination, Shanghai. Plans to extend it to Shanghai by 2005 also may be realized. But con-

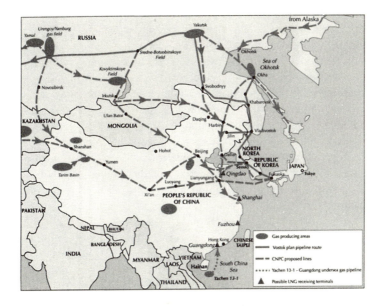

Figure 5.5 Proposed Pipeline Routs in Northeast Asia
Source: "China's Worldwide Quest for Energy," OECD/IEA Publication, 2000.

necting the grid northward or farther west, to Tarim and Karamay, join in Tuha, and then connect to a feed-in line from Qinghai and eventually on to Xian is at best a decade or more away.

Moreover, linking a Chinese gas grid to 3000+ km pipelines from Kazakhstan or Turkmenistan to Western China is a long-term proposition, conceivable no sooner than in a 2010–2015 time frame if commercial viability matters. Similarly, no matter how many letters of intent or memorandums of understanding are signed at Sino-Russian summits, building $10–$15 billion pipelines 5,000–6,000 km pipelines from Irkutsk through China to Korea and Japan (who would have to help finance such projects) are at best long-term propositions. An analysis of transport costs suggests that lengthy pipelines in excess of roughly 4,000 km are significantly more expensive than LNG: shipping LNG 6,500 km from Qatar to Japan is, for example, half the cost of piping gas 6,800 km from Novosibirsk in Russia to Shanghai. Moreover, new mobil LNG ships under development may significantly lower costs to gas consumers by avoiding the need to build regasification stations. Nonetheless, demand projections suggest that by 2020 China may need to import nearly 40 percent its gas requirements—presuming it meets ambitious domestic gas expansion targets.⁷⁴

For Southeastern China (Guangdong, Fujian, Jiangsu, Zhejiang, and Shanghai), where demand is large and growing and whose large cities are distant from coal and hydropower sources, LNG appears an attractive option. As analysts from the Pacific Northwest Laboratory have argued, "Compared with coal-fired plants, LNG fueled power plants would operate at much higher efficiencies, require half the initial capital cost, and create a fraction of the environmental damage."⁷⁵ A new generation of combined-cycle gas turbines makes gas—both pipeline and LNG—an increasingly attractive option. Compared to coal-fired power, these technologies can enable plants to be built in half the time, at roughly half the cost, operate with almost twice as much thermal efficiency; produce far less particularates, SO_2 and CO_2 emissions; and produce power at cheaper costs per kilowatt hour.⁷⁶ Beijing has begun to move forward on plans to build three LNG terminals in southern China that could begin operation in the 2002–2005 period, the first of which was approved in 1999. But at a time of growing deficit and unemployment fears, such plans compete in the near term with pipeline construction to distribute domestic gas.

Conclusion: Outline of an Energy Strategy

One can begin to discern the outlines of an emerging national energy strategy with different emphases on a regional basis: coal and domestic oil in Northern and Northeast China; hydropower in the Yangtze River delta; nuclear, Middle East oil imports, and LNG where demand is concentrated from Hong Kong along the coastal area up to Shanghai; an emphasis on gas in Western and Central China with a pipeline from Sichuan to Wuhan and gradually connecting both north and east to Shanghai as a national gas grid takes shape over the next generation. In southwestern China (Yunnan and perhaps Guizhou), domestic gas and tie-ins to a Southeast Asian grid may evolve. The pace and scope of such developments, however, will depend on the success of market-reforms discussed above and how plans are implemented from region to region, province to province. One factor is the degree to which China pursues uneconomic choices in a quest to substitute "diversity at any cost" for energy self-sufficiency. Two other "wild card" factors are the degree to which China's profound environmental problems shape energy decisions, and the unpredictable possibility of new energy technologies, such as fuel cells, which, in the 2010–2020 period could significantly alter the energy equation. In the near term, there will be substantial continuity in China's energy mix, changes in energy use invariably are a protracted process.

Whatever emphasis Beijing pursues, some energy realities are inevitable. Like the United States, China will become a major oil importer, by 2020 importing half to two-thirds of its oil, no less than 80 percent of it from the Middle East. Discussions with Chinese officials and intellectuals suggests that China is clearly uncomfortable with its petroleum dependency.[77] At present, most Middle East imports to China are from Yemen and Oman, China is expanding its energy relationship to Saudi Arabia, which would seem to offset its energy ties with Iran and Iraq in geopolitical terms. Certainly, Beijing is wary of Islam in general, confronting growing unrest among its own Moslem minorities, largely ethnic Turk Uighurs in Xinjiang province, and its ties to Iran and Iraq (discussed in chapter 4) appear principally commercial in nature.

While China may seek financial gain in its (diminishing) arms sales in the region, and has a geoeconomic interest in being engaged in the region, as one prominent Middle East specialist has argued, "China wants a peaceful, stable Middle East and to avoid entanglement in regional conflicts or crises. Rather than seeking a sphere of influence or strategic advantage, Beijing prefers to focus on trade and economic development."[78] This view is echoed by one well-connected Chinese energy analyst who claims, "China's goals in the region (like its penetration in the Middle East and Central Asia) are to enhance its oil interests rather than political or military expansion."[79]

The problem flowing from China's oil import dependence is not as Samuel Huntington has argued, that a "Confucian-Islamic connection" is emerging to challenge "Western interests, values and power."[80] There is simply little compelling evidence of such a trend to date. Rather, the question is whether over time China adjusts to importer status and becomes comfortable enough to rely on markets, long-term supply contracts, strategic petroleum reserves, and other economic measures for its energy security. In the near to mid-term this would mean remaining a free rider on the U.S. security role in the Gulf. The alternative would be for China to develop a competitive blue water navy, something that is at least 15 to 20 years over the horizon. But such a move would impede the sort of interstate cooperation required to manage energy needs, particularly oil and gas. For example, the development of gas requires substantial investment, be it in pipelines or in liquefaction plants or receiving stations. Gas is also based on long-term contracts, 30 to 40 years, to amortize investments. If much discussed Russian gas options are pursued this will require cooperation and financing from Japan and South Korea, and perhaps from international financial institutions, which might also fund Central Asian gas options. A security environment of heightened strategic competition, with naval arms races between China, India, and Japan, would be a not inconceivable outcome if Beijing aggressively sought to displace the United States with a large blue water navy. It is difficult to envision the sort of international economic cooperation necessary amid such a scenario.

One interesting expression of Chinese strategic thinking on energy was an article summarizing a report by a Chinese think tank published in *Ta Kung Pao,* a pro-PRC Hong Kong daily. Acknowledging that to the middle of the twenty-first century, China would be dependent on Middle East oil, the report said:

> China has no scheme in the Middle East for a global strategy. All it cares about is economic security. . . . In order to insure a safe supply of oil, China will seek to promote a long-term peace and stability in the Middle East, guarantee unimpeded passage the international oil routes and safeguard a stable and sustained international supply of oil from the Middle East, This is not in conflict with the Middle East policy of the United States.[81]

At the same time, however, the report also argues that "China should unite with Asian countries, particularly East Asian countries like Japan and South Korea . . . to jointly assume certain responsibilities in regard to promoting political and economic security and stability in the Middle East." It adds, "China should join with East Asian countries to establish an East Asian oil safe supply and reserve system."[82] Such thinking appears to reflect Beijing's discomfort with being a "free rider" on the U.S. Navy, which dominates the Persian Gulf sea-lanes. It also seems an acknowledgment of the reality that it will be at least two decades before China has the naval capacity to challenge U.S. naval dominance. In essence, the logic of the report's recommendations suggest a search for an interim path that reduces dependence on U.S. dominance.

Concerns—both in China and in the West—that China's anticipated oil demand may trigger either world oil shortages and/or price spikes—do not appear accurate. Even high-end projections are unlikely to trigger either world oil shortfalls or to spike prices. Using likely scenarios for Chinese growth, and presuming a conservative 1 percent annual growth in non-OPEC production to 2010—less than it grew in a low-price environment from 1988 to 1997—the probable outcome, as one authoritative analysis concluded, "is either a slight oversupply that could be relatively easily managed by OPEC or a larger surplus similar to that which drove prices down substantially in 1998."[83] In reality, it is likely that non-OPEC production, particularly both offshore West Africa, will grow more significantly, as discussed in chapter 2: the cost of finding and producing offshore oil has dropped precipitously. Moreover, OPEC at the beginning of 2000 was shutting in more than 5 million barrels a day, and Persian Gulf producers are actively pursuing new foreign investment in their respective oil sectors, raising the prospect of still larger increases in production. This suggests the IEA median case of 93.8 million b/pd and projections of 111 million b/pd world oil demand by 2020 will not necessarily mean steady price spirals, or prospects of oil shortages, as Chinese demand escalates.

Finally, another dimension of energy security, environmental degradation, is likely to be a growing factor in Chinese energy decision making. Environmental questions are directly linked to economic reform. Nearly two-thirds of China's primary energy consumption has been in the industrial sector. It is precisely the Maoist-era dinosaur heavy industry, much concentrated in the northern "rust belt," that is the most energy inefficient. China's iron and steel industry, for example, uses on average one-third more energy per ton of steel than U.S. industry.[84] Yet as reform has advanced, China has made some impressive gains in reducing its energy/GDP intensity. As Vaclav Smil has shown, between 1980 and 1995, the national average energy intensity shrunk from 0.7 kilograms of coal equivalent per 1 *Yuan* of GDP to 0.35 kg. This is a remarkable rate of decline, perhaps testimony to past inefficiency. But much of these gains have been the result of closing down old factories and upgrading others.[85]

To the degree that China's current push for reform and the competitive pressures from joining the WTO succeed in altering its economy, small, inefficient factories will shut down larger industry will be rationalized, and the net effect is likely to be new gains in energy efficiency. Moreover the degree to which China switches from coal-fired power plants to gas will also have a large impact on environmental quality. It was only in the mid-1990s that Beijing began to take seriously the dimensions of its environmental crisis. The degree to which environment becomes a factor and how much the physical intolerability of air and water pollution is reflected in China's political system is difficult to predict.

In sum, like China itself, its energy predicament and the policies fashioned to address it are a work in progress. China has only begin to make a host of difficult choices to shape its energy future. Some of these may also be revealed in its choices on military modernization. Perceptions of what best assures energy security, whether geopolitics or geoeconomics, predominates, the politics of national-provincial relations and tradeoffs between economic stability, market efficiency, environmental integrity are all elements in the matrix of China's energy choices. But China's emergence as a global energy player need not be disruptive to global stability or a negative factor in the world economy. Indeed, if managed appropriately, energy could be an important element fostering China's regional and global economic integration.

Chapter 6

India: The Lumbering Elephant

The conventional wisdom often draws a sharp distinction between India, "the world's largest democracy," and its neighbor, authoritarian, Leninist-ruled China. Of course, India's parliamentary democracy, institutionalized rule of law (however imperfect), and civil society stand in sharp contrast to China's Confucian-Leninist political system. No less distinctive are their respective histories and strategic cultures. Yet focusing on its perhaps surprising commonality with China helps illuminate India's considerable energy challenges, its dilemmas, how it hopes to address them, and its notions of energy security. Beyond their equally large, sprawling geography, huge and 70+ percent rural populations (total: India, 1 billion; China 1.25 billion), however, the two Asian behemoths have remarkably parallel characteristics: as ancient civilizations; victims of Western imperialism; and curious, if less immediately apparent, contemporary economic predicaments and energy challenges. India's economy reflects the Fabian socialist legacy of an economy that is similarly inefficient, bureaucratically inert, and thus far, slower to change than that of China. Both their economies are burdened by a similar legacy of autarchic development, state control, and economic nationalism that has biased policy, hampering modernization no less in India than in China—and looms as a hindrance to be transcended on the path to meeting their respective energy challenges.

Sino-Indian similarities are particularly striking in regard to most aspects of their energy predicaments—and their respective responses: the need to alter the energy resource mix; energy deficits; challenges of economic reform; geopolitical considerations shaping their energy situations; structural problems; and difficult energy choices ahead—with success in both cases largely dependent on reform. Consider:

- The majority of both Chinese and Indian commercial energy depends heavily on coal (China 67 percent, India roughly 60 percent) as both are the first and third largest producers with the third and fifth largest world reserves, respectively;
- Both are major oil producers, yet are net importers increasingly dependent on Persian Gulf crude for the majority of their oil (and in India's case, gas) as rising middle classes and transportation sectors point to sharply rising oil consumption;
- Both have considerable hydroelectric potential, still underutilized, but increasingly being realized;
- Both are expanding the use of natural gas (pipeline and LNG), that is the fastest growing energy source in both nations' energy mix;
- Both have ambitious nuclear energy plans, yet civil nuclear power programs which are faltering and producing barely 2 percent of their primary commercial energy;
- Both are major league polluters, with some of the world's most polluted cities who together, by 2020 may exceed the United States and Europe in SO_2 and CO_2 emissions yet have 10 (China) to 20 (India) times lower per capita energy use than the United States;
- Both have energy sectors dominated by bloated, state-owned monopolies in the process of reforming, and between them will require some $500 billion in capital investment for energy infrastructure over the next 10 to 15 years.

Though India's economy, roughly $430 billion GDP in 1999, is less than half the size of China's $950 billion economy, it has a larger middle class, estimated at roughly 150–180 million, though the number is a matter of some debate.[1] Unfortunately, it also has a similar, if not larger, bottom rung of some 250–300 million living in or near absolute poverty and a lingering caste system that impedes social mobility. It provides some perspective on Delhi's ability to cope with its vast energy and other problems to note that India's population has more than tripled over the past half century—from 300 million at independence in 1947, to one billion by 2000. While India's economy does not boast China's long-term near double-digit growth rate, it has been a more steady, buoyant performer than often appreciated. India's economy, grew at 5.4 percent from 1985 to 1995, and averaged just over 6 percent annual growth in the period 1996–99, with rates of 6 percent to 7 percent projected out to 2010 and 2020.² The uptick in growth is in part attributable to a process of deregulation (particularly toward foreign investment in the energy sector) and privatization that began to unfold in 1991, though it has been fitful and piecemeal in implementation.³

The new path embarked upon in 1991 marks a major—if gradual and phased—policy shift, toward market-oriented reform, constrained by residual tension from ingrained bureaucratic and political interests. This has also involved a reassessment of energy policies with an evolving strategy of using domestic resources more efficiently and attracting substantial foreign investment for urgently needed energy infrastructure and technology. Since the Nehru era at the start of independence, statist economic policies have been a dominant feature of India's political economy, beginning with the 1948 Industrial Policy Resolution that defined oil as one of six key industries to be run by the state. In the decades that followed there was a gradual nationalization of industry and development of the state sector, along with controls on foreign capital and a generally inhospitable environment for Western direct investment. Energy subsidies based on artificial prices over time became deeply entrenched in India's system of political patronage: politicians running on promises of free electricity have not been a highly unusual phenomena. Some of this began to generate rumblings of change in the late 1980s, as India saw dynamic East Asian growth take off, particularly as it saw China enter its second decade of double-digit growth, attract massive foreign direct investment, and begin to emerge as a major regional and global economic player. The end of the Cold War and demise of the Soviet Union (and loss of Soviet economic assistance) also helped reshape Indian politics, its geopolitical perspectives, views about foreign investment and the efficacy of its traditional economic policies.

The collapse of the USSR parallels the decline and disarray of the Congress Party, which lost its ruling majority after 1990 following the tragic assassinations of first, Indira Gandhi and subsequently Rajiv Gandhi. At the same time, there has been a steady growth, if rather sudden ascendancy, of the Bharatiya Janata Party (BJP) and of regional-based parties in a turbulent domestic political environment of shifting coalitions that has made for unstable governance. However, the September 1999 national elections, which gave the BJP-led coalition a solid working majority, appeared to mark a watershed: the first time over the past decade that a government had a real prospect of fulfilling its five-year term. Despite its Hindu nationalist cast, in regard to the economy in general and energy in particular, the views and policies of the BJP, led by Prime Minister Atali Bihari Vajpayee, have become more business-friendly. After major economic nationalist backlashes to proposed foreign investment, most notably the controversy in Maharashtra state that temporarily derailed a $3 billion Enron Corporation power project in 1995, Vajpayee appears committed to moving ahead with an agenda of deregulation and privatization.[4]

This steady erosion of its entrenched socialist mind-set and policies and consequent movement toward the global economy coincides with generational change in India (more than half the population is under the age of 24), leading one of Asia's most prominent economic newsweeklies to characterize the shift in views toward business and foreign investment as "India's Awakening."[5] There is a burgeoning entrepreneuralism, of which India's world-class software and computer industry is an oft-cited symbol, and growing consumerism, from both middle classes and aspiring middle classes. If urban Indians are buying air conditioners, motor scooters, autos, washing machines, and computers, rural Indian villages (rather than individuals) may be acquiring TVs, VCRs, and fans. But at the same time, only 80–85 percent of India's roughly 600,000 villages were wired for electricity. Moreover, commercial energy sources meet only 60 percent of India's needs (in large measure, catering to the 25–30 percent of the population that is urbanized) while the other 40 percent is comprised of traditional (e.g., firewood, organic wastes) and renewable fuel sources. Some 70 percent of India's population remains rural. This suggests the potential for more rapid urbanization—and rapid demand growth, manifested in more use of commercial fuels and more transport and residential sector demand over the next generation—if, as expected, India's economic growth trajectory continues.

India's Energy Challenges

Such demographic and economic trends begin to point to the magnitude of the energy challenges facing India over the next two decades. India's energy consumption is projected to more than double by 2020.[6] But already, electrical power generation, for example, is throttled by inadequate capacity and poor transmission and distribution, resulting in supply gaps in some areas (there is no unified national grid) 30 percent below demand, and nationally, estimated variously at 11 percent to 18 percent at peak periods. The result is constant power outages, something of a perpetual energy crisis, and India's most urgent immediate energy challenge.[7] Since 1950,

India's electricity demand has grown by an average annual rate of 8.8 percent, and is conservatively projected to grow in the 5 percent range over the next decade—the highest demand growth rate of any major country.[8] In barely a quarter century from 1970, a period of modest economic success, India's energy consumption roughly tripled.[9]

In its previous five-year plan (1992–97) the government met only 60 percent of its targeted increase in electrical capacity, less than 10 percent of it from private investment.[10] As India's growth has gradually accelerated, energy bottlenecks threaten to become a significant impediment to more robust economic expansion. By some estimates, power shortages may be costing 1–3 percent of national income.[11] India's dominant use of coal, much of which is high ash-content, is a major source of its substantial environmental problems. Even as India's consumption of petroleum products is set to skyrocket, its oil and gas production has been largely stagnant and its oil import bill rose to nearly $8 billion in 1999, roughly one-quarter of its export earnings.[12] Despite aggressive efforts to expand domestic exploration and production of oil and gas, oil imports are projected to increase roughly threefold to 3 -3.5 million b/pd by 2020, and its plans for expanding use of natural gas are in large measure dependent on foreign investment and imports.[13]

The policy challenges facing New Delhi were candidly outlined in a 1997 report of the Parliamentary Committee on Energy, which also stressed, "Energy is security; deficiencies in this critical sector compromise national security." The report then identified the array of problems:"The major issues in the energy sector are the absence of an integrated long-term policy; inefficiencies in energy supply and utilization; an unsustainable energy mix; acute scarcity of developmental capital; a lack of rational energy pricing; insufficient environment considerations ... these issues are all critically interdependent."[14]

The Indian government, which overwhelmingly dominates the energy sector, appears to have shaped the broad contours of a long-term approach (and has begun to fill in the details) designed to address such problems. It is a set of policies comprised of several major elements, the first—and perhaps most important—of which is to accelerate deregulation and privatization. This agenda includes reducing and gradually eliminating tariffs and subsidies on crude oil and petroleum products, decontrolling coal prices, and also making gas prices and power tariffs competitive. Yet only in April 1999 did a phased dismantling of India's Administered Pricing Mechanism begin. Though the government is accelerating implementation, many market distortions remain. Over the longer term it involves privatizing the bulk of the coal, oil, and power sectors. These steps are all aimed at increasing the amount of exploitable domestic energy resources. Another important component is to take the legislative, fiscal, and regulatory steps to create a substantially more attractive environment for foreign investors. This includes streamlining the license approval process for private power producers, offering more incentives for upstream oil and gas exploration, and promoting joint ventures, particularly for power generation. At the same time, New Delhi seeks to substantially improve the energy infrastructure with much-needed new refineries, urban gas transmission and distribution networks, a unified national grid, and improved transport with an ambitious north-south, east-west road program.

There are several key obstacles to realizing India's quest for stable, secure, and sustainable energy supplies at relatively low prices, not least, the self-imposed encumbrances of India's long-standing economic patterns. While these vary in kind and in degree from one energy source to another, they are essentially twofold in nature: the internal, problems of political/bureaucratic inertia and entrenched resistance to outward-looking liberalized trade and investment policies; and built-in constraints of geology and geopolitics, though the latter is ultimately malleable. One problem in enforcement, implementation, and integration of various measures, indeed of a comprehensive approach to energy security, according to one Indian analyst, is "that the energy sector is not the purview of a single government Ministry or Department of Energy."[15] Instead, there are a plethora of ministries and departments: of Petroleum and Natural Gas, Coal, Electric Power, Non-conventional energy resources, the Department of Atomic Energy, the Minister of Steel (coal imports). Then there is, for example, the Ministry of Surface Transport that oversees energy imports on state-owned ships.

This energy bureaucracy and the state-run enterprises that hold monopolies in the various industries in the energy sector tend to be mutually reinforcing, with the latter being a large component of the raison d'être of the former. For instance, public sector enterprises (PSEs) control upstream oil and gas, oil refining and marketing, and transport (pipelines) and distribution of gas. A host of regional-based subsidiaries of a major PSE, Coal India Limited, control the coal industry, and similarly, nine State Electricity Boards (SEBs) have controlled the generation and transmission of electricity. These government monopolies have subsidized kerosene, liquid petroleum gas, and perhaps most notoriously, electrical power, particularly to rural areas, (consumers pay significantly less than cost), a pattern entrenched in India's political system. This has left SEBs financially strapped (1998 losses: $3 billion), inefficient (average generating capacity utilized is barely 50 percent, with high transmission and distribution losses of 21 percent), and dependent on state governments for subsidies.[16] This in turn adds to India's national debt, with states comprising nearly 40 percent of India's total debt, which is a relatively high 10 percent of GDP. Moreover, privatizing PSEs would remove the rationale for the existence of corresponding ministries, which, as we have seen in China, have indeed, been abolished or streamlined as a result of energy sector reforms.

One instance of the worst of both worlds, where both policy flaws and geological fate are intertwined, is the use of coal. Like China, India depends on its vast coal reserves, some 77 billion short tons (7 percent of the world's proven reserves) and total ultimate recoverable coal resources estimated at some 220 billion short tons, for the bulk of its primary energy.[17] Coal accounts for roughly 55–60 percent of India's primary energy mix, while oil accounts for some 30 percent, hydro for 6 percent, natural gas for about 6 percent, and nuclear for about 2 percent.[18] (See Figure 6.1.) Also similar to China, industry accounts for more than 60 percent of India's commercial energy consumption. Moreover, from 1980 to 1996 industrial sector energy consumption grew more than threefold, from 2.7 quadrillion (quads) British thermal units (BTUs) to 8.6 quads. In comparison, the

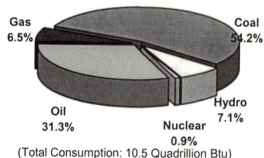

Figure 6.1 India's Energy Consumption by Fuel: 1995
Source: U.S. Energy Information Administration.

transportation sector roughly doubled from 1990 to 2000, while residential sector primary energy consumption jumped from 0.4 quads to 1.6 quads in the 1980–96 period.[19]

Coal is still the workhorse of India's energy production, with coal-fired thermal power plants providing 60 to 70 percent of the nation's electricity. Roughly 70 percent of India's coal consumption is used for power generation, producing about two-thirds of all carbon emissions.[20] Most of India's coal is concentrated in northern and northwest India—in Bihar, West Bengal, and Madhya Pradesh. This type of high ash-content coal has limited uses and requires India to import coking coal for steel production. Coal also means substantial transport costs on India's railways and is subject to strikes from the nation's powerful railway unions. Characteristic of India's energy sector, a state-owned conglomerate, Coal India Limited (CIL) acts as a kind of holding company, with eight regional subsidiaries accounting for some 90 percent of India's total coal output. CIL has many of the typical problems also evidenced in China's state energy companies: overstaffing, low productivity, inefficiency, and inadequate transportation infrastructure. In the mid-1990s Delhi began to deregulate the coal industry, liberalizing prices and distribution, and BJP proposals to allow the Indian private sector to develop new mines, with only minority interest of foreign investors, were being considered by the Parliament in late 1999.[21] CIL estimates that it needs some $4 billion in investment in the coal industry from 1998 to 2003, particularly foreign technology to aid in both underground and opencast mining. India has reached loan agreements with the World Bank and Japanese Export-Import Bank for $1.6 million to rehabilitate existing mines.[22]

But coal, of course, also is a prime culprit in India's serious environmental problems. Since 1980, India's economy has increased 2.5 times, but industrial pollution has quadrupled, while pollution from vehicles (85 percent of which use cheaper but more polluting diesel fuel) has increased eightfold.[23] According to the World Health Organization, New Delhi is one of the world's ten most polluted cities. But the prospective costs involved in remedying the situation suggest that replacing older coal-fired plants will be a difficult and protracted process, as will the introduction of clean coal technologies on a significant scale.

India Looks to Market Solutions

In light of this situation, India's policy response began to evolve with its 1991 economic reforms, initially by allowing private investment in the power sector, then tax holidays and easier access to 100 percent foreign equity projects. Over the course of its current five-year plan (1998–2002) India seeks to add 10,000 MW of power a year, with an estimated infrastructure cost of some $150 billion. India's president, K. R. Narayanan, bluntly told the Lok Sabha (India's Parliament), "The most obvious constraint to rapid economic growth today is the inadequacy of our infrastructure."[24] New Delhi estimates it will need 111,500 MW by 2007.[25] Indeed, the BJP-led government's "Power Vision 2010" agenda hopes to add a total of 80,000 MW of power over the 1998–2008 decade. To realize this rapid expansion, India has launched a series of reforms to improve plant load factor (now at less than 50 percent of capacity in many instances) and generate more private sector activity. It is also seeking to diversify its sources, increasingly looking to other, more environmentally compatible sources of electrical power, particularly large-scale hydroelectric projects and the expansion of natural gas. To achieve the required private and foreign investment, India has gradually expanded its menu of incentives, further altering its regulatory policies and tax policies to rapidly expand capacity. In 1998, it eased rules for foreign investment in its power sector, allowing automatic-approval 100 percent foreign equity in large-scale projects.

One measure of the shortcomings of previous government efforts to adequately address its energy challenges—and the priority the BJP administration is attaching to energy policy—is a recent study designed to clarify national goals and its reform agenda, "Hydrocarbon Vision 2025." The Vision study calls for India to drop limits on foreign ownership for investment in refineries, allowing 100 percent ownership. The study proposes phasing out petroleum subsidies in the 2003–2005 period, linking the price of natural gas to fully international prices (currently they are only 75 percent linked).[26] While past performance raises questions as to how effectively this latest round of reform pledges will be implemented, the renewed focus on deregulation suggests continued seriousness. Moreover, new pressures from a U.S.-India private sector-based Oil and Gas forum, created as one result from President Clinton's March 2000 visit to India, may also facilitate deregulation and privatization efforts.

The desperate need for growth in power infrastructure to match its economic growth has led India to look to private financing with government counterguarantees (central government covering default by SEBs) on eight "fast-track" power projects, though its efforts to implement attractive schemes have been mixed.[27] The national government is trying to encourage "mega-projects," defined as more than 1,000 MW thermal power, or more than 500 MW hydroelectric. While these do include several coal-fired plants, large-scale hydroelectric projects figure prominently in this new energy policy, as does liquefied natural gas (LNG) and also small scale (3 megawatts or less) projects for remote rural areas. India has also become one of the world's largest users of wind power, which along with solar energy is likely to be a niche energy source. India seeks to expand hydropower from roughly 25 percent of its 97 GW of electric generating capacity (world's sixth largest—more than the United Kingdom and France combined).

India has considerable hydroelectric potential, estimated at 84,000 MW, and the government has adopted policies to promote the development of hydroelectric power. Though some major dam projects have sparked controversy on environmental grounds, harnessing the Ganges, Brahmaputra, and their tributaries as well as other rivers can be an important source of renewable energy. In 1998–99 12 large-scale projects were approved, the locus of which was in north and northeast states, such as Manipur, Sikkim, Jammu, and Kashmir. Among the steps taken to facilitate increased hydropower is a planned tariff subsidy from which Delhi hopes to raise some $725 million annually. Other measures include an easing of regulations, exempting any hydroelectric facility up to 250 MW from having to seek technical or economic approval from the Central Electrical Authority.

Oil & Gas: The Geopolitical Dimension

But even more than hydroelectric power, India is looking to natural gas, particularly LNG, as a clean and cost-effective source of power for the future, one that can reduce the proportion of coal in its energy mix. Gas consumption was virtually nil before 1987, when India's last major new fields, offshore Bombay High (the source of 70 percent of India's gas) began to come online. Currently, India produces 23 billion cu. m/yr of natural gas, which has been the fastest growing energy source in India in the 1990s. It has nearly doubled from 1990, to 0.6 trillion cubic feet (tcf) in 1995, and is projected to double again to 1.2 tcf in 2000. In this, India reflects a broader new Asian trend to develop gas, indeed, over the next generation, cross-border and gradually, subregional (Southeast and Northeast Asia) energy grids as part of an energy security strategy that encompasses environmental concerns, though much of this involves both foreign direct investment and long-term dependence on imported fuel. (See Figure 6.2.)

Unlike the case of coal, hydroelectric and other renewables, or nuclear energy, the domestic-centered components of its energy mix where India

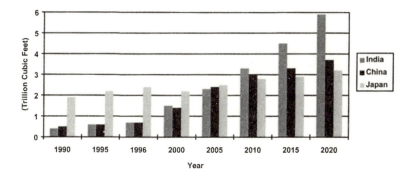

Figure 6.2 Natural Gas Consumption of India, China, and Japan, 1990–2020
Source: U.S. Energy Information Administration, "International Energy Outlook, 1998."

can gain some measure of its energy security largely by altering internal policies and practices, hydrocarbon energy sources necessarily add a geopolitical dimension. Indeed, in the case of India, its geography in relation to potential oil and gas sources makes for a particularly complex geopolitics of energy. India's endowments of gas as well as oil are significant but relatively modest, only .59 percent of the world's total proved reserves for both oil and gas. The magnitude of India's current and projected oil and gas needs mean that it must look to imports for the overwhelming majority of supply. The move to gas, particularly LNG imports, initially concentrated near ports in Western states (LNG terminals are also planned for Tamil Gnat on the southeast coast) to fill a growing portion of its power needs, is a product of the urgency of India's electricity deficit, its economic reforms, and growing receptivity to foreign investment. It is also a reflection of India's effort to secure stable, cost-competitive energy supplies, as well as its hope of diversifying energy sources.

India's oil and gas industry has been controlled by state-owned monopolies, overseen by the Ministry of Petroleum and Natural Gas. The Oil and Natural Gas Corporation (ONGC) controls the bulk of India's upstream oil and gas activities; Oil India Limited (OIL) controls upstream activities in northeastern areas; Indian Oil Corporation Limited (IOC) dominates India's refining and marketing, as well as imports of crude oil; and the Gas Authority of India Limited (GAIL) was formed by the government in 1984 and has been responsible for the transportation (e.g., pipelines), distribution, and marketing of natural gas in India. The refining sector, barely one-fifth the size of China's, is somewhat more diverse, though its shortfalls have led India to import large amounts of refined petroleum products over the past decade.

However, India has launched what on paper, at least, appears a major effort at making its oil and gas industry more competitive and more attractive to prospective foreign partners whose involvement will be one measure of the success of both India's reforms and its energy security, in the case of the downstream sector, deregulation, including tax holidays for new refineries, provisions to allow foreign firms to invest in refinery infrastructure, and an opening to foreign firms in the lubricants market, with the promise of opening up the retail marketing area over the next few years. The government has accelerated the dismantling of the administrative price mechanism for petroleum products. As one of a number of steps toward privatization, ONGC and OIL entered into an equity swap deal in the hope of melding the former's upstream dominance and the latter's downstream operations into more globally competitive, integrated multinationals. GAIL has also pursued some partial privatization, selling shares on the market. And in the upstream sector, the Indian government unveiled its New Exploration Licensing Policy (NELP) to facilitate foreign involvement in the exploration and production of oil and gas, granting equal treatment with domestic firms and tax holidays upon production, though results have been disappointing thus far.

One area where India's new market-oriented policies have begun to make large strides is in the development of gas, particularly LNG. New Delhi is investing heavily in infrastructure, the LNG terminals and pipelines to expand the use of gas in its current (1997–2002) five-year plan, and the U.S. Energy Information Agency (EIA) projects 8.6 percent average annual growth in gas consumption to 2020.[28] Although in the past India has never imported gas, its major state-owned firms, led by ONGC and GAIL, have formed a joint venture, Petronet LNG, to build LNG receiving terminals and gas pipeline grids and import LNG in partnership with foreign suppliers and investors.[29] India has approved six of some 14 proposed gas projects, and four large projects, all financed and arranged by Western firms, are under way. All the current projects are on India's west coast, mainly in Maharashtra and Gujarat on the Arabian Sea, and all are planning on long-term contracts from the Gulf (Oman, Abu Dhabi, Qatar, and Yemen). There are also plans to construct two gas import terminals on India's southeastern coast in Tamil Nadu and Andhra Pradesh.

Enron Corporation is on the leading edge of the new Indian push to

LNG, overcoming earlier political obstacles that led to postponement of the Maharashtra project. This project has broader significance for prospective foreign investment as a symbol of India's new openness. Enron is the majority partner (General Electric and Bechtel Corp. are also involved) in the $2.8 billion project to construct an import terminal to supply its 2,140 MW power plant at Dhabol, near Bombay in Maharashtra state. The first phase, involving a 740 MW naphtha-fired plant, started up in June 1999, with the second phase, 1,400 M of capacity, to be commissioned by the end of 2001, when the entire plant will be fueled on natural gas.[30] A consortium led by British Gas is to build an import terminal at Pipavav with imported LNG from Yemen to supply Gujarat; the French firm Total has initiated a facility at Trombay to supply gas to a local power plant; and a consortium led by Siemens (including Unocal) is also building a gas import plant at Ennore, near Madras.[31]

Indian firms have also initiated a number of gas projects on their own. GAIL plans to lay pipelines to inland towns and cities from the west coast, including a 550 km, $350 million gas pipeline in Gujarat for completion in 2003. India's Petronet consortium is arranging a 25-year supply deal with Rasgas of Qatar.[32] While some of this gas may be sold to the Dabhol power plant, Enron has worked out long-term supply arrangements for LNG from Oman and Abu Dhabi for much of its LNG. Enron has also signed a letter of intent with Malaysia LNG Tiga for an additional supply of 2.6 million tonnes annually, a 20-year commitment beginning in 2002.[33] India's artificial pricing and the lack of a coherent regulatory framework is a major impediment to an active gas market in India. While the government concedes the need to radically alter its fiscal regime by early 2000, its efforts were still in studying various schemes to do so. However, reform to put pricing on a fully market-competitive basis is unlikely before 2002–2003. Moreover, some of the import terminals, such as those planned by GAIL, and other infrastructure overlap with other Western-led efforts and may result in overcapacity.

Oil: The Middle East Connection

India's efforts to assure current and future oil supplies have also led to the same dependence on the Gulf/Middle East for oil that it appears to be developing for gas supplies. There is a touch of irony here in that India boasts the oldest oil industry in Asia, with its first (and continuously operated) field, at Digboi, near the Burmese border, producing since 1889.[34] Certainly India's oil industry has come a long way; some might argue, full circle: from the gradual nationalization of foreign holdings in the initial postindependence period to the creation of public sector oil companies and now it moves towards privatizing state-owned firms and a cordial opening to foreign investment in exploration and production with a sense of urgency.[35]

Until recently, India was able to supply more than half its oil requirements from domestic sources. But around 1993, roughly the same time as China became a net importer, India's oil imports exceeded 50 percent of its total consumption. At present, Indian production appears largely stagnant near a peak of 780,000 b/pd, while its oil imports for 1999 exceeded 1.1 million b/pd.[36] There have been no major finds in recent years, since major discoveries offshore at Bombay High in the mid-1980s. At present, India's reserve-to-production ratio is 14.5 years, substantially below the world average of 41 years.[37] Indeed, India's own five-year plan states that India will run out of oil reserves by 2012 unless new discoveries are made.[38] The government's New Exploration Licensing Policy (NELP) has attracted only modest interest from international investors since it was unveiled in late 1997. While some Indian optimists suggest that only a small portion of India's onshore and offshore basins have been explored, several officials of Western oil firms contacted by the author suggested no expectation that major oil fields were likely to be discovered.[39] In addition to doubts about oil potential, concerns that of 48 new blocks, Indian state-owned companies had reserved the best prospects for themselves, a host of continuing regulatory concerns, and obstacles to downstream activity (e.g., marketing oil products in India) all may help explain the limited Western response.

In sum, it is possible that advances in deep-water drilling technology may result in some modest new finds and that new investment in existing fields may prolong or modestly expand (India's 28 percent recovery rate is well below world averages) current production. But the bottom line is that Indian resources are unlikely to offer little alternative to oil imports to meet the bulk of India's burgeoning needs. Indian firms have been pursuing exploration activities abroad, in Central Asia, Oman, Vietnam, and Iraq, but thus far have not developed any significant new oil fields, except for some modest commercial quantities of gas in Vietnam.[40] All this points strongly to growing dependence on oil imports over the coming two decades, which, as noted above, may exceed 3.3 million b/pd by 2020. These trends also strongly suggest growing dependence on the Gulf/Middle East for much of its oil, which has traditionally been the case. Currently, the bulk of Indian oil imports are from Saudi Arabia, Kuwait, UAE, and Iran, and this is almost certainly likely to continue to be the locus of oil sources, despite Indian efforts to diversify supply.

Apart from its growing oil and gas ties to the Persian Gulf and its geographic proximity (also to Central Asia), India also has multifaceted social ties to the Gulf. There are some 2–2.5 million Indian workers and technicians in the Gulf states at any given time. These expatriate workers send remittances back to India in the range of $2 billion annually, partially offsetting the foreign exchange outflow from oil imports. These expatriate workers also have investments in the Gulf states, many of which were devastated a decade ago during the Gulf War, when an embarrassed Indian government floundered before eventually organizing an evacuation effort. Moreover, India is home to 130 million Moslems, the world's second largest Islamic population, mainly from the Sunni sect, although there are a significant minority of Shia Moslems. This factor has manifested itself in Indian foreign policy, as the government has sought to balance relations with the Gulf states as well as Iran. Moreover, in pursuit of stable, friendly bilateral relationships with Gulf states India has well-established military-to-military programs including joint exercises and exchanges with Saudi Arabia, Iran, Qatar, Oman, Israel, and prior to the Gulf war, with Iraq as well. But it is important to note that not withstanding these relationships, India has not sold significant arms or military technology to the Gulf states. This is consistent with India's general track record of restraint on the export of sensitive nuclear and missile-related technologies.

The Tyranny of Geopolitics

Regardless of the efficacy of India's reforms, India's hopes of diversifying supply as a means of enhancing energy security have been—and will continue to be—constrained by its political geography for the foreseeable future. India's geopolitical realities in regard to oil and gas resources, are in many respects, the reverse of many other Asian actors. For example, Japan has spent tens of billions of dollars over the past three decades searching for oil fields in the Middle East or pursuing oil and gas schemes with Russia in an effort to diversify supplies. More recently, as we have seen, China has also purchased oil fields from Sudan to Central Asia and entertained the idea of constructing massive pipelines that have little economic rational in the hope of diversifying supplies in the name of geopolitics and energy security.

But for India, the opposite is true: energy projects that have a compelling economic logic are stifled by the geopolitical realities of South and

Southwest Asia. Positioned on the Arabian Peninsula, only 885 miles separate India's westernmost bulge from Oman on the tip of the peninsula, a major supplier of oil and gas. Turkmenistan's enormous gas reserves are also within range of commercially viable pipelines to India. Only its land border with Pakistan physically separates India from Iran. And of course, northeast India shares a long border with gas-rich Bangladesh, whose independent national existence was obtained with Indian support in 1971. India has sought oil and gas relationships with all these actors in the region, only to fall prey to impediments, largely political in nature with the exception of Oman, where technical/economic problems have sidelined plans for an Oman-Bombay pipeline.

Some prime examples of this dilemma involve Turkmenistan gas, Iranian oil and gas, and Bangladeshi gas. In the case of Turkmenistan gas, a Unocal-led consortium had laid the legal and financial groundwork for building a 1,400 km pipeline from Turkmenistan through Afghanistan to Multan, Pakistan, with an additional 600 km connecting link to New Delhi. By the end of 1995 agreements were reached between Turkmenistan, Pakistan, and the Unocal group to build the pipeline, which could send some 1.6 billion cu. ft. of gas a day into Pakistan and some portion to India. Construction was due to begin by the end of 1998, with initial deliveries to Pakistan in 2001. Yet in November 1998, continued instability and conflict in Taliban-led Afghanistan led Unocal to cancel the project. Another eventual obstacle may have been the pathology of Indo-Pakistani rivalry. Though Pakistani officials had publicly indicated support for the Indian leg of the pipeline as a "win-win proposition," prior to its cancellation, it is unclear whether Pakistan would in fact have permitted the pipeline to be extended to India.[41] It would have entailed a 20–year commitment and created a degree of energy dependency amongst two rival states between whom tensions are persistent and armed conflict is a periodic reality. Such cooperation, indeed, regional free trade, has been the premise of the South Asian Association for Regional Cooperation (SAARC), a grouping inspired by ASEAN in Southeast Asia (see chapter 8). But SAARC has been hostage to Indo-Pakistani rivalry. In any case, following Unocal's withdrawal from the project, Turkmenistan, Afghanistan and Pakistan agreed to continue the project up to the Pakistan end point—pending the finding of a new majority partner to replace Unocal who could organize and obtain financing for the project. The India leg appeared to drop off.

Skepticism about the prospects for Indo-Pakistani energy cooperation is reinforced by Pakistan's reluctance to cooperate with a number of Indo-

Iranian energy schemes in recent years. Mehdi Hashemi, then Iran's deputy minister for gas, explained during a July 1999 visit to India, "Our government has been in talks with the Indian government for the last 7 years.... Iran considers India to be one of its biggest and near markets. It would be in India's interests to buy gas from one source which is near and never-ending rather than from several sources in far-off countries."[42] Iran, with 16 percent of the world's proven gas reserves, could become a long-term source of competitively priced gas. India and Iran have discussed a range of means of delivery: overland or underwater pipelines, overland train routes, and LNG by sea. All the prospective routes require passing through Pakistani territory or territorial waters, but a pipeline would be up to 20 percent cheaper than LNG according to Iranian officials quoted in *Oil & Gas Journal*.[43] Though at various points in different fora Pakistani officials had in principle agreed to permit a pipeline route (which could provide handsome tariffs and transit fees) and pledged not to disrupt the gas flow, no such permission has been forthcoming.[44] Indeed, 1999 saw renewed military clashes over Kashmir and a lingering climate of tension into the new century.

Both the Afghanistan and Central Asia issues highlight a second dimension of Indo-Pakistani competition: Pakistan has historically had close ties to certain Islamic states, and India's prospective energy ties to the region may be viewed as an effort to compete for influence. But this is secondary to the centerpiece of Indo-Pakistani rivalry symbolized by the Kashmir question. This predicament also highlights the double-edged nature of energy cooperation. On the one hand, a network of energy ties could give both India and Pakistan a stake in stability and serve as a deterrent to future conflict in their ongoing confrontation, ostensibly over Kashmir (in reality, it is a deeper existential dispute that goes to the core of each state's identity). Yet these energy issues could also add another layer of strategic competition in the shifting geopolitics of the region.

If India's reforms accelerate economic takeoff and a mind-set of growing prosperity, with neighbors viewed more as trading partners than threats, that could alter regional dynamics. Pakistan could follow a parallel course of reform and modernization, though to suggest such a scenario in mid-2000 is to invite charges of naïveté. It is also possible that Islamabad might feel still more fragile and threatened by a more powerful emerging India and pursue a course of confrontation. In light of the emotional, indeed, visceral nature of Indo-Pakistani rivalry, it is difficult to envision substantial energy cooperation from Pakistan or for that matter an Indian willingness to allow any of its energy lifeblood to become vulnerable to Pakistan.

In this regard, it may be worth noting that only in late 1999 did Egypt and Israel agree to construct a pipeline to send Egyptian gas to Israel—more than two decades after Anwar Sadat's bold initiatives led to an Israeli-Egyptian peace, albeit a cold peace.

Perhaps even more frustrating for India, and certainly the oddest geopolitical impediment to what would appear a natural economic relationship, is the case of neighboring Bangladesh. One of the world's poorest nations with few natural resources, Bangladesh is believed to have considerable natural gas reserves. Though proved reserves thus far are a modest, but significant, 11 trillion cu. ft., promising geology and lack of seismic work and drilling have led major Western energy companies to believe that Bangladeshi gas reserves are substantially larger, in the range of 50 trillion cu. ft., which would make Dacca a major producer. Some estimates are as high as 80 trillion cu. ft.[45] For a country with few resources other than jute and cheap labor, gas could be Bangladesh's great hope for economic development.

India, particularly the neighboring market of Calcutta, and the still larger market in Delhi, are natural markets, whose proximity to Bangladesh would offer a competitive price advantage to gas from the Gulf or Central Asia. Yet a host of self-imposed obstacles have stifled even the exploration and production of Bangladeshi gas, despite strong interest from many major firms including Shell, Unocal, Occidental, and Halliburton.[46] Both GAIL and ONGC are interested in exploring Bangladeshi gas reserves, and Unocal has interest in developing a recently discovered field on the Bangladesh/Burma border for export to Calcutta, with the hope that selling to Unocal rather than directly to India might be more palatable to Dacca. Lack of expertise on complex technical issues—how to offer exploration blocks, how to manage the bidding process for licensing rights, developing its own gas industry and a framework for exports—has slowed the development of gas production. Bangladeshi national politics and pathologies add another dimension to the problem of cross-border energy cooperation.

Bangladeshi fears that economic integration with India would disadvantage its domestic industry and, more broadly, increase Indian regional domination, have led to a general apprehension about economic ties to India. In regard to natural gas, a combination of 1970s "North-South" ideological fear of foreign multinationals and concerns about squandering natural resources they consider their birthright for Indian advantage have led to a wariness about gas export development schemes. Yet it is such export prospects that will make investment more attractive to foreign investors. In addition, neither of the two major political parties, the Awami League and the Bangladesh National Party (BNP), has ever reconciled itself to the other's legitimacy, and thus the party out of power strongly attacks the policies of the party in power. For example, Prime Minister Sheikh Hassina's 1998 Ganges water-sharing accord with India was assailed by the BNP. Given the potential for lifting its economy, Dacca's handling of natural gas remains puzzling. "Why the Bangladeshis are so reluctant to sell gas to India is one of the great mysteries," confessed one energy analyst for a prominent Western consulting firm.[47]

Energy Security

Over the long term, some of these geopolitical obstacles will eventually yield to the dynamic of what appear natural economic patterns, as this analysis suggests is the case. Yet even were all the obstacles discussed above to melt away (and Indo-Pakistani rivalry will not disappear in the foreseeable future) the above discussed prospective oil and gas sources would only provide India a modicum of diversity of supply. While achieving more diverse supply patterns would reduce its sense of vulnerability to disruption, India's future is one where two-thirds or three-fourths of its oil will be imported and similar proportions of its natural gas. That is of course, less foreign oil dependence than that of Japan or South Korea, and roughly the same degree of dependence that the United States will face by 2020. But as we have seen, apart from potential short-term disruptions, this need not necessarily generate a sense of energy insecurity, as global supplies are more than adequate. It will, however, require long-term contractual arrangements with Gulf/Middle East oil producers and stockpiling precautions to hedge against short-term disruptions. This approach, its market-oriented reforms, particularly, its growing receptivity to foreign investment, and its LNG policies suggest in broad strategic terms that a reliance on global markets and cross-border investment is the core of India's approach to energy security.

Yet this market-based approach to energy questions, one in which economic interdependence and global markets are key, is often not reflected in the discussion of such questions by many Indian policy intellectuals. Indeed, when discussing its energy situation the vocabulary and logic frequently employed by Indian national security analysts tends to be old-fashioned nineteenth-century military neomercantilism. In an essay on Gulf-India security ties, Sujit Dutta, a generally thoughtful and well-informed analyst at a prominent New Delhi think-tank argues: "India's oil dependence on the Gulf has a direct bearing on the country's military planning and preparations. The Indian navy and air force are critically important in terms of protection of supply lanes. Accordingly, India is engaged in ongoing efforts to strengthen the armed forces to perform this mission."[48]

This is hardly an isolated view. Another Indian analyst argues that "Pakistan's strategic location astride the SLOCs [Sea Lines of Communication] will necessitate sufficient naval and air capabilities to ensure the security of crude oil and natural gas to India." To ensure the freedom of navigation required for oil and gas flows, he argues that attaining aircraft carrier capability is "a necessity, in order to seriously ensure the security of SLOCS in the Persian Gulf and the Arabian Sea, critical to India."[49] And India's Foreign Minister, Jaswant Singh, in an essay seeking to make the case for India as a nuclear weapons state, argues, "A more powerful India will help balance and connect the oil-rich Gulf region and the rapidly industrializing countries of Southeast Asia."[50] Why this is the case is not clear. Singh points to India's "extensive energy import requirements" and to missile and nuclear proliferation in the Gulf to rationalize the notion that nuclear weapons make India a Great Power and this somehow provides a strategic balance.

Yet would such military approaches really enhance India's energy security? In fact, the globalization of oil markets offers more stability of supply to India than either a blue water navy or a nuclear arsenal. Citing Pakistan's "strategic location along the SLOCs" is a bit far-fetched. While it has invested in significant naval (and antisubmarine) forces, Islamabad is not known for its naval prowess. It lacks both the capacity to disrupt the flow of tanker traffic, an extreme act of war that would provoke an immediate military response from Delhi. Moreover, such a move would almost certainly trigger a response from the U.S. Navy, for whom freedom of navigation is an inviolate principle, and likely censure from the U.N. Security Council. Moreover, such an act would have global consequences that would not only harm Pakistani interests and those of its allies such as China but impact global oil markets writ large.

Dr. Dutta, while speaking of Indian missions to guard oil supplies, concludes realistically in the same essay, "It is not likely that the United States will abandon its role or be replaced as the guarantor of Gulf security anytime soon." This would seem to mitigate any urgency for an expansive Indian naval mission. He also argues that a breakthrough in U.S.-Iranian relations would serve Indian interests and perhaps allow the evolution toward a new multilateral approach, "the development of creative and cooperative approaches to Gulf security."[51] In any case, India's naval budget allocation steadily diminished during the 1990s, with 12 of 40 main combatant ships and one of two small aircraft carriers due for decommissioning by 2001.[52] India has been considering the purchase of a Russian carrier and several frigates to replace those being decommissioned, and has tested a sea-launched nuclear-capable missile based on Russian technology. Thus far, however, India has not displayed a concentrated effort at naval modernization that would reflect an imminent threat to vital SLOCs. Indeed, its naval budget has steadily shrunk in recent years. A study for the Pentagon's internal think tank, the Office of Net Assessments, observed, "India's military priorities reflect a growing concern with its northern borders and internal security issues. The strategic importance of India's naval prowess in the Indian Ocean does not receive the same attention or resources . . ."[53]

In fact, if India did pursue an expansive blue water navy replete with aircraft carriers, destroyers, and submarines with full-blown air and sea force projection capabilities, it would almost certainly result in less, not more security. This would be a classic case of what political scientists call a "security dilemma," where moves by one actor to increase its security inadvertently creates a sense of insecurity on the part of another actor. Such an Indian military agenda would spark a naval arms race with China and in all likelihood, Japan, which in 2000 began reviewing Article 9 of its postwar "peace constitution," already has the most sophisticated and capable hi-tech air and naval forces in the Pacific, and would likely expand its maritime mission.

Such a scenario or a variation of it may unfold in the first quarter of the twenty-first century. But in actuality, it would have little to do with energy per se. As discussed in chapter 4, disruption of oil tanker traffic, particularly given the fungible nature of oil in a global market, would spike the price of oil for all importing nations. It is difficult to envision an oil embargo as was done to Japan in the late 1930s and early 1940s apart from a scenario of a protracted World War II type conflict. Otherwise, short-term disruptions can be managed through stockpiles and supply diversity. Energy may provide a rationalization for such military programs. As was the case between Germany and Britain at the end of the nineteenth century—documented in Robert K. Massie's magisterial book, *Dreadnought*—military competition can have a psychology and momentum related less to particular security threats than to great power ambitions and/or generalized fears.

Such logic may help explain the apparent disconnect between India's energy policies, whose trajectory clearly points to market-based solutions, and defense arguments ostensibly rationalized on the basis of energy vulnerabilities. In fact, behind talk of blue water navies is a similar logic to that inherent in India's nuclear policies. In a definitive book on India's nuclear weapons program, *India's Nuclear Bomb,* George Perkovich makes a compelling case that the development of India's nuclear weapons is the product of an influential strategic elite scientific-technical "strategic enclave" combined with India's notions of national identity and great power aspirations. "India's national identity," writes Perkovich, "is constructed around the determination to be an independent great state that transcends its colonial past and is morally superior to its colonizers and the dominant states of the international system."[54] Thus India's steadfast rejection of the nuclear nonproliferation treaty (NPT) as reflecting an unfair and unequal system, creating, as Foreign Minister Singh argues, "nuclear apartheid."

For India, nuclear power, beginning in the immediate post-independence period, and nuclear weapons *capability* were symbols of modernity, of great power status. Prior to 1998, India was able to cling to its moral posture by maintaining a nuclear option, but stopping short of obtaining operational nuclear weapons while advocating nuclear disarmament. But the fact that it was 24 years after 1974, when India demonstrated nuclear capability with its "peaceful nuclear explosion" before it declared itself to be a full-fledged weapons state, suggests that its effort to attain nuclear weapons was not driven primarily by a clear and present threat. Moreover, the Indian military has been kept out of the decision process on nuclear weapons, largely confined to the Prime Minister and the scientific-technical "strategic enclave."

This may help explain why two years after its nuclear tests, India still had not devised a nuclear doctrine, or established a command and control system or decided on the architecture of operationally deployed weapons. These realities suggest motivations other than existential survival at play in India's nuclear gambit. Once these decisions are taken, the Indian military necessarily acquires a major role in regard to nuclear weapons. India's claims that threats from China spurred its weapons decision, while understandable, given the trauma resulting from the 1962 India-China border war, and lingering concerns about Chinese behavior, remain unpersuasive given the limited nature of Sino-Indian disputes. Undoubtedly there is a psychological undercurrent of resentment and perhaps envy at the great power treatment lavished on China. But among other things, China's economy (twice the size of India's) is far more integrated into the global economic system, growing over 9 percent annually for two decades. Certainly the near-term effect of its nuclear weapons program is to justify Pakistan's nuclear program, which neutralizes India's overwhelming conventional military advantage and gives Islamabad a shield behind which it can pinprick India on Kashmir.

Thus, while to be sure, there is a context of security concerns surrounding India's nuclear weapons program, it appears driven by other factors. Certainly, India jealously guards its version of the U.S. "Monroe Doctrine," as the preeminent power in South Asia, and views developments, such as Chinese presence and influence in Burma, particularly in the Coco Islands in the Sea of Andaman as signs of potential encroachment.[55] India's nuclear weapons program, like talk of blue water navies, may be better explained more as a psycho-political pathology bound up in notions of national prestige and identity as a great power than it is as a function of a plan to safeguard energy security. Nuclear weapons are viewed as a means to obtain India's rightful place as a great Hindu civilization and major power.

This chapter has conspicuously not emphasized India's policies in regard to nuclear power. This is because, Indian arguments to the contrary notwithstanding, nuclear power is and will likely remain tangential as an energy source. Currently, nuclear power provides just under 2 percent of India's total energy. There has also not been a clear line separating India's civilian power and weapons program. One former U.S. official says, "Almost all of its reactors have had both civilian and weapons-related roles at one time or another."[56]

The historic symbolism of India's nuclear program continues as part of its national myth. Thus, a senior Indian economic official told a Harvard audience in late 1999, "Nuclear energy has to be an important component of our overall strategy to meet our rapidly growing energy demand."[57] In light of India's relatively large national debt, the five- to seven-year lead time involved in building nuclear reactors, and the large upfront capital cost, its nuclear power program is likely to expand only modestly to 2020, though it is considering purchase of a Russian-built reactor and also has four partially built reactors. In addition, there are limitations on the transfer of nuclear technology from nuclear weapons states due to nonproliferation concerns. Moreover, India has troubling safety problems at many of its ten functioning reactors at six nuclear plants. Though veiled in secrecy by an ominous Official Secrets Act, accounts have surfaced documenting a number of radioactive leaks, fires, and at least one near meltdown at Indian reactors. The crusading work of Adinarayana Gopalakrishnan, a nuclear engineer and former chair of India's Atomic Energy Regulatory Board, has revealed some of these problems. Gopalakrishnan's contract was not renewed after he submitted a 300-page safety report detailing 95 safety issues requiring urgent action.[58] Says a Western official familiar with India's nuclear program, "In most other countries, many of the reactors in India would be shut down and decommissioned." Moreover, the performance at India's nuclear plants is poor, on average, roughly half the load factor of OECD nation reactors.[59]

Clearly, nuclear power figures only marginally in India's future energy mix. As we have seen, both government and private sector are laying the foundation for natural gas playing a major role in providing future electrical power. As is the case in China, the implementation of the economic reforms India has embarked upon will be a determining factor shaping India's energy security. India appears determined to become a full-fledged great power in the twenty-first century. Its capacity to do so will require it to become a multidimensional global actor—economic, technological, civilizational, as well as military. How India responds to its energy challenges will be illustrative of the degree to which it achieves such status.

Chapter 7

Japan/Korea: Whither Asia's Mature Consumers?

No countries better epitomize the fear and trepidation of energy pessimists than Japan and Korea. Only the United States imports more oil (about 10 m b/pd) than Japan and Korea who together import nearly eight million barrels of oil a day. Japanese and Korean perpetual—and growing—energy deficits, highlight the concern that the future portends a world of scarcity creating new vulnerabilities in Northeast Asia, one where military conflict is sparked by a struggle over control of diminishing resources. It has been argued that Northeast Asia, according to one prominent analyst, is "a region where energy and security are locked in an unusually tight and fateful embrace."[1] But are the energy strategies of Japan and/or the Republic of Korea (ROK, herein referred to as South Korea) premised on the notion that the future is a contemporary version of East Asia's experience in the 1930s and 1940s, a desperate quest for energy *lebensraum?*

While memories of oil embargoes past still shape the "resource-poor, island nation" thinking of Japanese policy-makers, they are gradually fading as traditional Japanese scarcity fears are tempered by less harsh and more manageable recent experiences. In fact, a casual observer could be forgiven for wondering why Japan and many other Asian states appear particularly concerned about energy security in the twenty-first century. Recall the stark reality that Japan and South Korea, two of the region's most stunning success stories, are bereft of any significant hydrocarbon resources. Nary a drop of oil. Yet at the center of the breathtaking Asian success stories witnessed over the past century are the Japanese phoenix rising from the ashes of World War II, and then of South Korea transforming itself from a per capita income equal to that of the Congo in 1960 into a globally competitive OECD nation with a $10,000 per capita GDP by 1996.

The Japanese and Korean post–World War II experiences suggest that deficits of oil or other raw materials—however challenging as policy choices—may be secondary factors affecting both economic growth and national security. Annual oil import bills in the $50 billion range have not prevented Japan from running annual trade surpluses well over $100 billion. Certainly the twin oil shocks of 1973–74 and 1979–80, though politically and economically traumatic at the time, seem in retrospect to have been only bumps in the road for both the Japanese and Korean economic juggernauts. Indeed, it was in the 1990s, when oil prices were at historic lows, that Japan's economy entered a period of stagnation. Japan and South Korea, respectively, the world's second and twelfth largest economies, are among the world's largest consumers of oil and gas, more than 99 percent of it imported. With barely 5 percent of the region's population between them, Japan (5.5 m b/pd) and South Korea (2.1 m b/pd) account for some 35 percent of the roughly 20 million b/pd of oil consumed by Asia-Pacific countries in 1999.[2]

Japan, with a population nearly three times that of South Korea, has a $4.5 trillion economy, more than eight times larger, and one that is more mature, less energy-intensive, and more energy efficient than that of Korea. Both are pursuing policies of deregulation of their respective energy sectors with varying degrees of vigor, processes that in both cases appear incremental in implementation. Though they differ in size, in national psychology, and geopolitical perspective, Japan and Korea have remarkably similar energy patterns and comparable policy responses that each in their own way reflects their respective national character shaped by history and geography:

- both have similar energy mixes, with oil accounting for 55 percent of total primary consumption in both nations, coal 16 percent in the case of Japan, 20 percent for South Korea; nuclear power 14 percent and 12 percent respectively, natural gas, 12 percent for both, and hydro 3.8 percent and 5 percent (see Figure 7.1);
- both Japan and South Korea import over 99 percent of their oil, 80 percent of it from the Middle East;
- Japan and South Korea both are increasingly looking to liquefied natural gas (LNG) as a clean energy source and together, account for nearly 75 percent of the world's imports of LNG;
- both have invested heavily in nuclear power, which supplies 35 percent of Japan's electricity; 40 percent of Korea's.[3]

- both are treaty allies of the United States, host forward-deployed U.S. forces, and view American security guarantees (not least, of the sea-lanes) as a critical component of their national security.

There are, however, some important differences as well beyond those of size and proportion of energy requirements. Japan is a more advanced, postindustrial economy that has impressively wrung much energy inefficiency out of its economy. From 1975 to 1994 the amount of energy needed per $1 million of Japan's GDP was reduced from 210 tons of oil equivalent (toe) to 110 toe.[4] It is focused on energy conservation, and no OECD nation has made a more serious effort to meet the emission reduction targets of the Kyoto Protocol on Global Warming than Japan. In contrast, the Korean economy is more dependent on energy-intensive industries (steel, shipbuilding, petrochemicals, autos, cement), more focused on rapid growth. South Korea by contrast has put much less emphasis on conservation. Thus, South Korea requires almost twice as much energy as Japan for every $1 million of manufacturing output. And one very large, indeed, incalculable difference is the enormous question mark hanging over South Korea's future: the fate of economically failing North Korea and the potential costs of reunification.

Another intriguing difference—particularly in light of their similar situations and similar policies—is that of respective Japanese and Korean perceptions and national psychologies in regard to energy. Though South Korea is, if anything, more vulnerable than Japan, it is relatively sanguine about energy security, while the Japanese are consumed with energy security concerns. In part, this reflects the fact that Japan conceives of itself as a major power, while Korea is self-consciously a middle power of more modest ambition. Where Japan frantically seeks to overcome its basic predicament, Korea has a certain toughness and confidence that it can somehow cope. "The difference," says a senior Korean diplomat, "is that we think of ourselves as a small continental power surrounded by giants and will find a way to get by, while the Japanese have the mentality of an isolated island."[5]

Japan's Energy Imperative

Far more than any other Asian nation, Japan, Asia's largest oil market, has been obsessed with the question of energy security for much of the past century. Securing access to oil, 60 percent of which was imported from the United States by Japan, was more than a minor causal factor behind Tokyo's

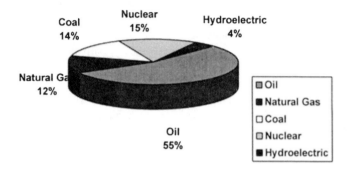

Total Consumption = 21.28 Quadrillion Btu

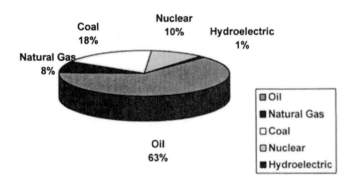

Total Consumption = 7.46 Quadrillion Btu

Figure 7.1 Energy Consumption by Fuel in Japan and South Korea, 1997
Source: U.S. Energy Information Administration, "International Energy Annual, 1998"

1930s colonial expansion in Asia and in the Pacific War. The U.S. oil embargo in 1941 cut off the bulk of Japan's oil. This led Japan to strike at Pearl Harbor, as it then viewed conflict with the United States as inevitable. The notion of Japan as a resource-poor, isolated archipelago off continental Asia has become a veritable idée fixe in the Japanese psyche. Geography and an absence of natural resources, however, only partially explain this phenomenon. Japan's approach to energy security (and the priority accorded to it) also reflects a worldview shaped by its traumatic historical experience in the oil era, first (self-imposed) during the period of imperial expansion in the 1930s and 1940s, and more recently as victim during the oil shocks of 1973–74, and again in 1979–80. Yet despite all Japan's efforts to alter its energy situation, its dependence on outside sources of energy remains over 80 percent, among the highest of advanced industrialized nations.

This sense of vulnerability and fragile dependence on outside forces animates Tokyo's energy policies and permeates much of its foreign policy in pursuit of the chimera of self-reliance. Indeed, there seems a lingering, quixotic belief, however unstated, that Japan's national will, properly directed, can somehow gain control over its energy fate. In the 1930s this was manifested in its imperial expansion, seizing the resources of China and Indonesia to fuel Japanese industry and a Tokyo-centered regional order. In the post–World War II period, and particularly since 1973, such national psychology illuminates why some of Japan's most flawed and costly industrial policies and extraordinarily tight government-business arrangements are in the energy sector.[6]

Over the past decade, however, favorable external developments on the supply front and the appreciation of the yen since the September 1985 "Plaza Accord" have relaxed energy fears somewhat. The amount of Japan's GDP spent on oil imports has declined ninefold since 1980, and by 1998, the yen cost of oil had dropped nearly 80 percent from 1980.[7] The lack of either disruptions or prolonged price spikes resulting from the 1990–91 Gulf War began to suggest to Japanese energy managers that world oil markets might be far less vulnerable than was the case in the 1970s. The growth of non-OPEC production and significant rises in both global production and reserves have meant low prices, reduced further by the strong yen. Furthermore, the demise of the Soviet Union opened up potential new oil and gas markets in Russia and the former Soviet republics.

This more benign energy supply picture has not, however, substantially altered the trajectory of Japanese energy policies. Officially, the goal of Japan's energy policy is described as the "3 E's": simultaneously achieving energy security, economic growth, and environmental protection. But the protracted stagnation of the Japanese economy has complicated its energy agenda. As discussed in chapter 4, Japan's economy, despite nearly $1 trillion in stimulus packages since 1992, has been stagnant, stuck in a deflationary cycle. Japan entered the new century technically in recession, with two consecutive quarters of negative growth at the end of 1999 and an estimated 0.6 percent growth for the year.[8] Its banking system holds nearly $1 trillion in bad debt, and government debt is now twice as large in proportion to GDP as the U.S. debt was at its worst in the 1980s.

In short, as now widely recognized in Japan, the "development state" political economy has become an impediment to Japan regaining its global competitiveness. This was conceded with remarkable candor by no less than Prime Minister Obuchi's "Commission on Japan's Goals in the Twenty-first Century," which concluded: "The vested interests and social conventions that have grown up during the course of Japan's development since the Meiji era (1868–1912) in accordance with the "catch up and overtake" model have ossified society and the economy and leached Japan's vitality."[9]

In fact, a gradual, if painful, process of economic restructuring has been under way since the mid-1990s. Industry accounts for roughly half of Japan's energy consumption. Energy prices significantly higher than in the United States or Europe, resulting from the heavily controlled energy sector, became a burden on already troubled Japanese business. Only toward the end of the 1990s did mounting internal pressures lead Japan's tightly regulated energy markets to begin to open up slowly to market forces. Deregulation of electric power and the downstream petroleum sector (distribution, refining) is gradually proceeding, aimed at increasing competition among domestic providers of energy and cost-effectiveness for business and consumers. In regard to energy demand, this analysis is based on a median between the IEA, "business-as-usual" projection for Japan of 1.8 percent demand growth to 2020, which is slightly higher than MITI's business-as-usual projection of 1.1 percent demand growth to 2010.[10]

The major elements of Japan's current energy policy—from choice of energy mix and sources of supply to conservation and stockpiling—trace their origins in large measure to the trauma of the 1973–74 oil shock, which resulted in 1974 becoming the first year since World War II that Japan saw negative growth. At the time, Japan's rapidly industrializing economy was highly dependent on oil, which comprised 77 percent of its primary energy needs. Though Japan imported 44 percent of its oil from the Persian Gulf states, the notion that it was enmeshed in the Middle East crisis and Six-Day War was not on Tokyo's radar screen.[11] Cheap oil—$3.31

a barrel in September 1973—was a given. But by October 1974, it nearly quadrupled to $11.47 a barrel.[12] But that was only the beginning. Suddenly, as an ally of the United States, Japan, as did Western Europe, found itself on the OPEC "unfriendly" list. Japanese diplomats and businessmen visiting Saudi Arabia and other Arab Gulf states were told in no uncertain terms, even neutrality was not good enough if they wanted to keep their oil flowing: Japan must support the Arab position in the Middle East.

In mid-November, then Secretary of State Henry Kissinger made an eleventh-hour effort to persuade Japan to align itself with the United States on the Arab-Israeli conflict. Days later, some European countries who had publicly backed the Arab side were excluded from oil supply cutbacks. That development, combined with Japan's obsessesiveness about resources, led Tokyo to tilt to the side of oil, its first major break with the United States in the postwar era. At the time, Prime Minister Kakuei Tanaka was a leading force in Japan's energy *zoku* (special interest groups). He responded swiftly to the pleas of nervous Japanese businessmen and MITI bureaucrats: on November 22, 1973, Japan issued a statement unambiguously backing the Arab position on the Middle East. Japan's move paid off, as it was spared from OPEC cutbacks made that December.

This oil shock–induced policy shift was the harbinger of a new era of Japanese "resource diplomacy." Diplomats and businessmen fanned out over the Middle East bearing aid, loans, proposals for joint ventures, and not least, looking for oil fields for which they could obtain production rights. Since in the new OPEC-centered oil world, major oil companies couldn't deliver, Japan would go directly to the source to gain assured supplies and cultivate the suppliers. An assertive melding of industrial policy (e.g., state sponsored global exploration and production activities) and foreign policy has since been an enduring feature of Japan's post-oil-shock approach to energy security. However, whether such efforts have actually contributed significantly to either energy security or economic growth is another matter.

Nonetheless, the new government-directed emphasis on Japanese firms acquiring equities abroad for exploration and production as a means of securing supplies and reducing vulnerability to future oil crises was but one pillar of a new set of energy policies adopted by Japan. "The first oil crisis, says a 1999 MITI energy policy paper, "prompted the Japanese government not only to formulate various emergency measures . . . but also to change the basic philosophy of its energy policy."[13] The result was a post-1973 energy policy with five major elements to achieve the "3E's": reducing oil dependency; diversifying non-oil energy supplies; securing supplies by building petroleum reserves via exploration and production; promoting energy conservation; and, accelerating research and development into new forms of energy. The second oil shock triggered by the 1979 Iranian revolution drove the price of imported crude oil to $33.46 a barrel by June 1980. This second price spike both reinforced the sense of urgency driving these evolving policy currents and gave new impetus to the development of alternative energy sources that, at such skyrocketing price levels, appeared competitive. It also lead to another rift with the United States, as Japan refused to isolate Iran. Though Tokyo maintained its oil supply relationship, under U.S. pressure it did subsequently suspend most other economic activities in the 1990s.

One by-product of the renewed emphasis on assuring energy supplies and developing alternatives was an augmentation of the already impressive power wielded by the "resource lobby" in Japanese policymaking. An accompanying rise was seen in the power and influence of those in MITI, the Foreign Ministry, the Agency for Natural Resources and Energy (ANRE), the Science and Technology Agency (STA), the Ministry of Transport, and other agencies relating to energy security. These forces, however, which often tend to act as bureaucratic fiefdoms, are sometimes prone to competition depending on the issues (e.g., gas versus nuclear or coal). Various agencies, along with particular *keiretsu* energy cross-holdings (e.g., refiners, oil importers, retailers), trade associations, and politicians can and have been impediments to deregulation in their respective energy sectors.

Oil Dilemmas

While Japan's policy response to the twin oil shocks led to a rapid expansion of nuclear power and to increasing reliance on natural gas, its efforts to remove oil from the center of its energy picture have met with only very limited success over the past generation. Japan is the world's second largest consumer of oil but has virtually no oil reserves of its own (60 million barrels). Oil has been reduced from 77 percent of Japan's total primary energy consumption in 1973 to 55 percent in 2000. Tokyo seeks to reduce the role of oil further to 47 percent by 2010. Its economy as a whole has reduced its dependence on imports for primary energy only from 89 percent to 82 percent since 1973.[14] One measure of the decline in dependence of its economy on oil is the fact that Japan's current oil consumption of 5.5 million b/pd is nearly the same amount as that of 1974, while its economy has more than doubled in size since then. This also reflects increased Japanese energy efficiency and success at conservation resulting from a series of measures (e.g., taxes, incentives for investment in energy-saving technology) adopted during the 1970s.

Less successful have been the Japanese government-directed efforts to enhance energy security by subsidizing the overseas activities of Japanese oil companies. This began in 1967, when MITI established the Japan National Oil Company (JNOC) for exploration and production. Since then, through a variety of investment programs, loans, and loan guarantees, JNOC has bankrolled more than two dozen Japanese oil companies searching for oil in East Asia, the Russian Far East and Central Asia, the Middle East, and Latin America. JNOC has provided financial assistance to more than 1,100 wildcat wells and 800 delineation wells around the world.[15]

In the case of China, JNOC had a 50 percent share in ten Japanese companies that were created from 1978 to 1988 to develop offshore oil in Bohai Bay, Beibu Bay, and in the Pearl River basin, as well as companies exploring in the Tarim Basin. In addition, at least $10 billion in Japanese government loans to China since 1979 have been for energy-related projects. Yet China accounts for less than 2 percent of the overseas oil production of Japanese firms, and several of the companies set up for exploration and production in China have gone bankrupt.[16] One of these, the Japan China Oil Development Corporation, is among three in a financial morass so severe that in 1999 MITI proposed a financial rehabilitation plan to liquidate them. Japan's energy connections with China are also motivated by a larger strategic goal of accommodating China's burgeoning energy needs with the aim of mitigating potential Chinese aggressiveness and strategic competition for resources.

The most successful Japanese exploration and production operations have been those of the Arabian Oil Company (AOC), a Japanese firm that for two decades operated the offshore section of the Saudi-Kuwaiti Neutral Zone. The Khafji and Hout concessions produce 300,000 b/pd of crude oil—roughly half of the total overseas production by Japanese firms. But the Saudi drilling rights for side of the field expired in February 2000, and Japan failed to extend the concession. This also puts at risk the Kuwaiti side, the concession for which expires in 2003.[17] Outside the Gulf, Mitsui owns a 15 percent share of the Caspian Kur Dashi oil field also owned by the Azerbaijan State Oil Company, which JNOC has said it would help finance. Moreover, at least four other Japanese companies have purchased shares of another Caspian field, Mitsubishi has a 10 percent share of the AIOC consortium in Azerbaijan, and Japanese firms also hold a one-seventh stake in the Kazakh field of Kashagan. Mitsubishi is also developing four large fields in Vietnam. One of Japan's most promising prospects is the Campos Basin in Brazil, where Mitsubishi and Itocho are part of a consortium. Despite these far-flung government-subsidized activities, 82 percent of its oil is imported from the Gulf/Middle East.[18] The UAE and Saudi Arabia account for 49 percent of Japanese imports, followed by Iran (10 percent), Qatar (7.8 percent), and the rest of the Gulf states (Kuwait, Oman, Neutral Zone), which account for another 17 percent. Asia Pacific countries supply only about 10 percent of Japan's oil imports.[19] Thus, oil—largely from the Gulf—remains Japan's primary energy source, however less dependent on oil its economy is than was the case during the first oil shock. Tokyo's goals for reducing the role of oil in its current energy mix are modest: it seeks to decrease oil to 47 percent of its total primary energy consumption by 2010, with natural gas and nuclear power correspondingly increasing. In regard to assurance of supply, Japan's goal for more than two decades has been to have overseas production comprise 30 percent of its total supply. Prior to the expiration of its Neutral Zone concession in early 2000, the sum total of Japanese exploration and production amounted to about 600,000 b/pd, or roughly 13 percent of its total imports. The expired concession was roughly half of the total Japanese overseas production. But it may lose all of its Neutral Zone concessions in the near future, when the remaining concession comes up for renewal in 2002. Unless unexpectedly large fields in Central Asia or Brazil are discovered, there is a possibility that even with the array of overseas prospects, Japanese firms' net production may not be significantly higher than recent production.

This raises the question as to whether JNOC-sponsored overseas ventures are, as then MITI chief Mitsuo Horiuchi told a Diet (Japan's parliament) inquiry in 1998, "essential for guaranteeing our nation's energy security." Since its inception, JNOC may have spent as much as $41 billion in its quest for more energy security. Its own figures (1.73 trillion yen by the end of fiscal 1996 to 266 development project companies) may be significantly understated. In fact, one published source cites a total of $30 billion through fiscal 1990. Several well-placed senior Japanese officials cited a figure of $41 billion through 1999 to the author, which would roughly correspond to the higher number, though I have not been able to document that figure.[20] Understatement of financial problems has been a frequent feature in Japanese economic controversies that dotted its political landscape in the 1990s—its $1 trillion bank debt was initially said to be $200 billion. Few would be shocked were JNOC's spending even higher. In any case, it is not surprising that JNOC has come under fire from the Diet over the past several years. One Diet member pointed out that the heads of nine disbanded oil companies funded by JNOC were former MITI bureaucrats. "There's been information," said the irate legislator, "to the effect that by expounding on the necessity of independently developed crude oil, MITI has been able to take tax money not under the control of the Finance Ministry and set up project companies and to create two layers of plush jobs for its former bureaucrats."[21]

Whether JNOC's activities have been a sincere effort or a bureaucratic boondoggle, the net result is, at best, a modest contribution to Japan's energy security at considerable cost. JNOC is a classic example of the downside of Japanese industrial policy. One can only speculate what the results of Japanese overseas oil and gas exploration might have been if its firms were tested by market forces. JNOC's operations received a scathing critique in a 1999 report done for MITI by the U.S. consulting group Booz-Allen & Hamilton. The report, part of a MITI oil policy review due for completion in 2000, said JNOC had made a number of bad management decisions such as failing to monitor the performance of projects, investing too much in exploration rather than acquired reserves, and failing to invest strategically. The report called for major reform including consolidating the number of Japanese exploration companies from more than 25 to 3, curbing government subsidies dramatically.[22]

Placed in the larger context of Japan's energy dependence, having equity stakes covering 13 percent of its daily needs may provide more of a psychological sense of security than a physical cushion in the event of another oil shock. Indeed, a far more effective shock absorber has been Japan's strategic petroleum reserve. In response to the 1973–74 oil shocks the advanced industrialized nations set up the International Energy Agency (IEA). Part of the IEA program was a stockpiling/crisis-sharing scheme requiring each member to establish 90 days of reserves. At present, Japan has 156 days' worth of oil in its stockpile. Given that the most likely danger to the flow of Japan's energy supplies is a short-term disruption (e.g., revolution in Iran or Saudi Arabia), Japan's large stockpile may be its best assurance against a threat to supplies.

Gas and Coal: Hydrocarbon Competitors

Japan is no less dependent on foreign sources for natural gas than it is for oil. Despite frequent discussion of—and elaborate proposals for—gas pipelines from the Asian mainland or Russian Far East to Japan, thus far, virtually all Japan's natural gas is in the form of liquefied natural gas (LNG). LNG, however, is a very capital-intensive process, requiring gas liquefaction plants and regasification receiving terminals to unload the LNG from tankers as upfront costs. Japan is the world's largest importer of LNG, accounting for 61 percent of global demand, though gas only comprises 11 percent of its total primary energy consumption. Some 48 million tons of gas is used annually, primarily in power plants to generate electricity. Japan, as is the case with the rest of Asia, uses less than half the proportion of gas in its energy mix than does the United States and Europe. Yet gas has been the fastest growing energy source globally over the past two decades, and this is true in Asia as well, where gas increased 18 percent annually from 1975 to 1995, albeit from a low base.[23]

For Japan, there are multiple and increasing virtues to natural gas. Gas is a clean burning fuel, and in light of Japan's effort to cut CO_2 emissions by 6 percent from 1990 levels to meet Kyoto targets, it is an increasingly attractive alternative to oil and coal. The geopolitics of its LNG are another attractive feature. In contrast to the situation with oil, the majority of Japan's gas supplies are from Southeast Asia and Australia—40 percent from Indonesia alone. Together with Malaysia, Brunei, and Australia, Asia-Pacific countries supply more than 80 percent of Japan's LNG (Abu Dhabi, Qatar, and Oman are its other suppliers). This energy interdependence forms a synergy with a larger Japanese foreign policy goal: playing a leading role in an integrated East Asian community. This is evident in Japan's efforts to cultivate ASEAN nations, most prominently Indonesia, the largest recipient of Japanese Official Development Assistance (ODA). The cumulative total of Japanese ODA to Indonesia is nearly 4 trillion yen, plus $900 million in emergency loans in response to the Asian economic crisis under its Miyazawa Initiative.[24]

In addition, Indonesia is only one of several venues where JNOC has been actively involved in exploration and production of natural gas. These include a stake in the ARCO-led consortium developing the Tangguh field in West Irian and the Exxon-led consortium in the $40 billion project to develop gas on Natuna Island. JNOC is also involved in a number of projects in the Russian Far East, with a 50 percent stake in SODECO, a consortium developing oil and gas in Sakhalin I and II, and also in gas projects in Yakhusk and Irkutsk. JNOC vice president Yasuhiko Wada has identified Russian gas as well as exploration in the Caspian as priorities in JNOC's investment strategy.[25] That prospective large new sources of natural gas is in the Russian Far East (especially Sakhalin) in close proximity to Japan, is yet another of its geopolitical virtues.

Much of the hoped for gas in new frontiers of exploration in Russia and Central Asia would not be in the form of LNG, but rather, delivered by lengthy pipelines. There are a plethora of grandiose plans, promoted by the powerful Japan National Pipeline Research Society whose 44 corporate members include the major utilities such as Tokyo Gas and Tokyo Electric Power Corp. (TEPCO), as well as companies like Mitsubishi, who would be involved in pipeline construction. Their supporters, including prominent ex-officials such as former Foreign Minister Taro Nakayama, have proposed forging an "Asia-Pacific Energy Community." There are variations on this idea, but the concept is to build a Northeast Asian gas grid connecting Irkutsk through Mongolia, into China, and on to Korea (in some plans) and Japan. This would, this vision holds, eventually be connected to a Southeast Asian grid reaching into southwestern China.

While it is easy to draw lines on a map, building pipelines in adverse permafrost conditions, undersea, and through deserts and forests is a complicated and expensive business. Moreover, transporting gas via long-distance pipelines (e.g., 3,000+ km) tends to be far more costly than LNG.[26] Absent a persuasive commercial logic, obtaining financing for multibillion-dollar projects may be highly problematic. As previously mentioned, various shorter segments of pipelines will likely be constructed in the 2000–2005 and 2005–2010 period (e.g., Wuhan-Sichuan-Shanghai, Sakhalin-Hoikkaido, Indonesia-Singapore). But a host of obstacles—from uncertainty about adequate supply to financing problems to difficulties in doing business with Russia—suggest such hopes will be more a pipedream, a fanciful vision, than a pipeline for the next two decades. The one international gas venture that by most indications makes economic sense—and is likely to be developed by about 2010—is Sakhalin gas, either via a Sakhalin-Hokkaido pipeline, or an LNG train from Sakhalin. Unlike other much-discussed Russian gas pipeline projects, Sakhalin is relatively close to Japan, some 600 miles from Hokkaido.

Japan has still further impediments to expanding either LNG or pipeline gas stemming largely from its highly inefficient system of refining, distribution, and marketing. Japan boasts the dubious distinction of being one of the only major gas-consuming nations with no domestic gas transmission or distribution network. LNG is imported at 20 unconnected terminals; roughly half is used in power plants next to marine terminals. City gas is trucked over urban roads to substations of city gas companies. Smaller distribution companies use propane. Moreover, as the *Oil & Gas Journal* points out, "less than 10 percent of urban Japan has any gas distribution system. Even the largest city gas utilities—Tokyo Gas, Osaka Gas, and Nagoya Gas—have distribution mains limited to 50km from the nearest LNG terminal."[27] The result is that end-users pay LNG prices more than three times higher than those in the United States or Europe, impeding the use of gas by businesses or for residential heating. This situation benefits Japan's nine large vertically integrated utilities, who are able to purchase in bulk, long-term (usually 20–30 years) supply contracts, with prices managed by MITI "guidance."

These inefficiencies have spurred deregulation efforts, as in a stagnant economy, business and consumers have chafed at bearing the burden of these inefficiencies. MITI began to put in place energy reforms aimed not at supply but rather at cost and efficiency. Reforms started in 1995 and 1996 began to inject a measure of competition into the distribution and marketing of gas, and even more since 1997, of petroleum. These reforms also allowed power utilities to purchase electricity on the wholesale market, opening the door to independent power producers (IPPs) and loosening the firm control wielded by the nine major utilities. But for deregulation to create an environment leading to the increased gas use requires promotion of competitive wholesale gas supply, but perhaps most importantly, a domestic pipeline grid to bring gas into Japan and trunk lines to distribute it. However, Japan's hugely indebted economy and continued economic stagnation make it difficult for Tokyo to invest in such infrastructure as public work programs, as might otherwise be the case. Beyond the financial difficulties, the transport and shipping companies as well as local gas distributors likely to lose out are vested interests that Tokyo would have to confront. This suggests it will be a gradual, piecemeal process over the coming 10 to 15 years, limiting the expansion of LNG demand as well as for pipeline gas. MITI's own projection has natural gas use increasing only slightly to 13 percent of primary energy supply by 2010.

Coal

The need for such investments and the long-term nature of LNG commitments have created short-term tension between LNG and coal. Japan is also the world's largest importer of steam coal for thermal power generation and coking coal for steel production. Altogether, Japan accounts for 28 percent of world coal imports. Of the 973 billion kilowatt hours of electricity generated by Japan in 1997, 64 percent was from thermal (oil, gas, coal). Japan imports more than 95 percent of its coal, about 130 million tons annually, most from Australia, Canada, China, and Indonesia. Coal accounts for 16.4 percent of Japan's primary energy supply, more than natural gas's 11.4 percent. The LNG-coal tension can be discerned in the discrepancies between power companies, that claim coal use will grow by 2.1 percent per year, and MITI which sees no need to rise above 1996 levels.[28] Indeed,

MITI projections show coal diminishing to 14.9 percent of primary energy supply, reflecting government efforts to reduce CO_2 emissions.

It remains to be seen what the impact of deregulation will be on coal demand. Japan's electricity prices are the highest in the OECD. MITI has revised its system for setting rates, and Japan's cabinet approved a plan aimed at reaching globally competitive electric power rates. But a number of utilities are less interested in Kyoto than in saving money, and in uncertain economic times they are looking to increase the use of coal to fill short-term needs, rather than getting locked into 10– or 20–year commitments if they renew or purchase more long-term LNG contracts.

Nuclear Quandaries

For Tokyo, nuclear power has long been the supreme symbol of its dream of energy independence. Well before the Kyoto Protocol, nuclear power was the darling of the Japanese bureaucracy. The scientific possibility of recycling spent nuclear fuel in a closed fuel cycle was a source of fascination for Japanese scientists in the period after World War II. In 1967, as Japan geared up its first commercial power reactors, the Japan Atomic Energy Commission (JAEC) decided to pursue the most controversial aspect of Japan's nuclear program, the reprocessing of plutonium and fast breeder reactors to attain a closed nuclear fuel cycle. However, the problematic situation of natural gas expansion and the still growing use of coal has put a larger premium yet on nuclear power in regard to meeting its environmental goals and reducing dependence on imported sources of energy. Japan's nuclear power programs predate the 1973–74 oil shock; its first commercial Light Water Reactors (LWRs) were ordered in 1965 and 1966, respectively. By 1973, nuclear power comprised just 3 percent of Japan's primary energy supply.

In the aftermath of the first oil shock, Japan began to place a larger emphasis on nuclear power as a key alternative to oil. Over the following fifteen years, nuclear power quadrupled to over 12 percent of Japan's primary energy supply. Japan has become the third largest producer of nuclear energy after the United States and France, with 52 reactors producing 45 gigawatts capacity. Currently, nuclear energy provides about 36 percent of Japan's electricity, and the government and nuclear industry hopes to build 16 to 20 more reactors by 2010 to increase nuclear's share of electricity production to 45 percent.[29] There are four new reactors either under construction or planned, three at existing nuclear plants. But there has been a sea-change in Japanese public attitudes about nuclear power in the 1990s.

There is a certain irony in this, as nuclear safety technology in LWRs has improved substantially since the 1970s when the nuclear industry was plagued by frequent technical problems and shutdowns.

A series of nuclear accidents, beginning in 1995—all at facilities involved in fuel cycle activities—has had the cumulative effect of calling into question the dominant view that the benefits in energy security from nuclear power far outweighed any risks. In an August 1999 poll conducted by the Prime Minister's office a majority expressed concerns about nuclear power, with 70 percent expressing fear of possible accidents; 49 percent opposed further development of nuclear power or felt Japan should abandon nuclear power.[30] These views were expressed *before* the worst nuclear accident in Japan's history occurred at the Tokaimura plant just 70 miles northeast of Tokyo. It was the first time a reactor went critical, with a chain reaction continuing for 20 hours. The mishap was the result of management cutting corners, with three workers mistakenly placing 35 pounds of enriched uranium into a container designed to handle only 4.8 pounds. One worker died, three were exposed to extreme amounts of radiation, 69 people received some exposure to radiation, and the town had to be evacuated. The accident raised questions about safety regulation, emergency management, and stoked fears of nuclear plants near urban areas.

It would be an exaggeration to call the accident Japan's Three Mile Island or Chernobyl. The Tokaimura accident rated a 4 (on a 1–7 scale) on the International Atomic Energy Agency's (IAEA) disaster scale. The meltdown of a reactor core at Three Mile Island got a 5; Chernobyl, a 7. But the Tokaimura accident may prove to have a similar impact on the fate of Japan's nuclear power industry as Three Mile Island did on that of the U.S. industry: destroying public support for further expansion. The Tokaimura episode follows a previous incident at the same complex in 1997, when a fire and explosion occurred that was then the most serious nuclear accident Japan had seen. And that followed a December 1995 accident at the Monju experimental fast breeder reactor where a large leak of sodium coolant led to an indefinite shutdown of the facility. Beyond the event itself, the political impact of a government coverup of the event and scandal (in which the official in charge of investigating the accident committed suicide) eroded public trust.

Following the Monju affair, Tokaimura may be a harbinger of the beginning of the end of Japan's understandable dream of reprocessing. After less than six months in operation Monju was shut down indefinitely, and the larger planned commercial reprocessing and breeder reactors, such as the Rokkasho-Mura breeder reactor, have been further postponed into at least mid-decade.³¹ Further postponements are quite likely, as several government commissions have been reassessing Japan's entire nuclear fuel cycle program. More recently, Japan has postponed plans to use mixed oxide fuel (MOX) as fuel in LWRs as a means of reducing stockpiles of plutonium it is accumulating from spent fuel reprocessed in Europe. All told, Japan has spent at least $31 billion on the back end of the fuel cycle (reprocessing) over the past three decades, according to several nuclear industry officials.³² Yet achieving the fuel cycle remains technically difficult and makes little economic sense in a world awash in plutonium and uranium from dismantled U.S. and Russian nuclear weapons.

Japan's accumulation of plutonium highlights the inescapable intersection of energy and security that nuclear power represents. With the capacity to build nuclear weapons if it chose, Tokyo is viewed as a "virtual" nuclear power by China and others in the region, though American extended deterrence has led Japan to refrain from any such proliferation. Yet both publicly and privately, Japanese officials have debated the nuclear option. One prominent example is a 1969 policy planning study subsequently leaked to *Mainichi Shimbun,* a leading national daily, which argued that Japan "should keep the economic and technical potential for the production of nuclear weapons, while seeing to it that Japan will not be interfered with in this regard."³³

In any case, in the near future, the lack of public acceptance makes expansion of Japan's nuclear power industry problematic at best.³⁴ Expressing a view not infrequently heard after the Tokaimura accident, one longtime Japanese nuclear industry official told the author, "I am doubtful that we will be able to build more than a handful of new nuclear reactors, and even then, only at existing plants."³⁵ What does this mean for Japan's future energy policies? Based on low-growth forecasts, Japan would need 5.9 m b/pd in 2010. Without any additional nuclear power by 2010, Japan would require an additional 1.17 m b/pd in oil to offset that energy deficit—if it substituted oil for all of the shortfall, according to calculations done by the Baker Institute. If it substituted natural gas for all the shortfall it would require 186.65 mil cu. mt./pd. Or in a third scenario, if the shortfall were replaced with 30 percent oil and 70 percent gas, Japan would need an additional 352,241 b/pd and 130.65 mil cu. mt/pd.³⁶

While such modifications in Japanese oil and/or gas needs would require adjustments in Japan's energy policies, the magnitude of oil and/or gas involved would not necessarily contribute to significant energy insecurity on Japan's part or for other actors in the region. Given the role of nuclear energy in electric power generation, more aggressive pursuit of expansion of natural gas would seem the most feasible alternative. But this raises some difficult financial problems. In Japan's current economic situation, the necessary substantial investment in domestic infrastructure to increase residential and commercial use of gas appears problematic.

Japan's Energy Future

Whether or not Tokyo must factor in replacement sources for nuclear power, the trajectory of its approach to energy security would not likely need to be altered significantly over the next decade. In broad terms, it presumes the maintenance of the U.S.-Japan security alliance, and U.S. control of the Sealanes of Communication (SLOCs). A study done for the Pentagon's internal "think tank," the Office of Net Assessments, obtained by the author concluded that Japan's "energy security planning is unfocused; badly fractured among many ministries; possesses no evident military element; and is overly optimistic about averting crisis based on the belief that market forces always will prevail."[37] That, it might be added, plus the U.S. Pacific fleet (not to mention Japan's own considerable air and naval capabilities) may offer confidence on SLOC security.

Japan's effort to diversify both its energy mix and its dependence on Gulf/Middle East oil will almost certainly continue. That effort is focused on Japan's "Eurasian diplomacy," first articulated by then Prime Minister Ryutaro Hashimoto in a July 1997 policy speech that presaged an important policy shift. After Japan initially maintaining its historic ambiguity toward Moscow in the aftermath of the Cold War in the early 1990s, a new set of circumstances began to change Japanese views around 1995–96. Not least was a growing concern about the meaning of the rise of China for Japan. This helped spur a rethinking of Japan's Russia policy, which, apart from historic distrust, was impeded by a long-standing territorial dispute over a group of four islands, the Northern Territories, claimed by Japan but seized by the USSR at the end of World War II. Renewed interest in Russian natural gas resources helped prompt a reevaluation. Finally, there was excitement over prospective new oil and gas resources in Central Asia touted by the United States.

Tokyo shifted its Russia policy, calling for "balanced expansion" of ties to Moscow. This meant a lower profile for the Northern Territories dispute in Russo-Japanese relations, and a more patient approach. This new approach has a complex logic in which energy plays a central role in geopolitical calculations. Japan seeks better ties to Russia in order to balance China. At the same time, development of energy resources is motivated not only to benefit Japan but to aid China's need for energy, for fear that shortages would lead to more aggressive Chinese behavior. Thus, Japanese companies, as mentioned above, have invested in two major Sakhalin projects, are exploring Irkutsk pipeline prospects, and are also partners in major oil and gas consortia in the Caspian Basin in Central Asia.

Clearly, the payoff from these efforts is in most cases at least a decade away. Japan's Eurasia strategy is long-term in nature. Large-scale pipeline gas from Russia is not imminent. Moreover, as discussed in chapter 3, the Caspian region is unlikely to produce the magnitude of oil that Japan had hoped as it geared up its diplomacy and its checkbook with aid and loans to Azerbaijan, Turkmenistan, and Kazakhstan.

As the Pentagon report concluded, Japan is relatively sanguine about its energy future. Beyond its diversification efforts, the development of new energy-efficient technologies is also viewed as key. As we have seen, Japanese auto companies are on the cutting edge of post–internal combustion engine technology. Toyota has been selling more than 2,000 Prius hybrid cars each month in Japan. While the commercial viability of these new auto technologies is yet to be proven, this is clearly the direction of the auto industry. It is a direction that looking out to 2015–2020 will begin to significantly reduce oil demand.

Korean Contrast

In many respects, the energy policy of South Korea is a scaled-down version of Japan's. If anything, South Korea has greater vulnerabilities than Japan, with some 95 percent of its primary energy demand met by foreign energy supplies. With no domestic oil reserves, South Korea is the world's fourth largest oil importer, with energy imports accounting for roughly 20 percent of its total import bill—more than twice that of Japan.[38] Prior to the 1997 economic recession, oil accounted for 63 percent of South Korea's primary energy. The 1998 recession reduced that to 54.2 percent, but a surprisingly robust 10 percent economic growth in 1999 appears to have restored both consumption levels and the percentage of oil in its energy mix (to the 60 percent range).[39] While Korea still has a host of serious financial and structural economic problems that were exposed by the 1997 financial crisis and are only slowly being addressed, this analysis assumes moderate growth (4–5 percent) over the coming decade.

Like Japan, it has sought—with limited success—to diversify sources of supply, particularly for oil and natural gas. The other basic elements of its energy policy are: to build larger strategic petroleum reserves; increase use of natural gas and nuclear energy; and increase energy efficiency and conservation. It is also looking to the energy frontier of the Russian Far East and the hope of energy cooperation in Northeast Asia to realize prospective oil and gas resources that could alter its supply patterns.

One difference between the Korean and Japanese responses to the two oil shocks was that South Korea expanded the use of coal as an alternative to OPEC oil dependency, particularly after the 1979–80 shock. By 1985, coal accounted for 39 percent of primary energy demand before environmental concerns and low oil prices reduced the role of coal in its energy mix nearly by half, to about 20 percent at present. In its moderate business-as-usual scenario, the Tokyo-based Asia-Pacific Energy Research Center (APERC) projects that coal will continue to account for roughly 20–23 percent of South Korea's energy mix to 2010.[40]

It is worth underscoring that despite immediate hardship at the time, the two oil shocks of the 1970s had little impact on South Korea's economic miracle. The energy growth pattern of Korea, which reached OECD income levels of just over $10,000 per capita by 1996 (when it became the world's sixth largest consumer of oil) may help illuminate future demand growth elsewhere in Asia. From 1975 to 1992, South Korea tripled its per capita energy consumption, while averaging about 8 percent growth annually. Its oil demand quadrupled from 1985 to 1995, reflecting rapid structural change in the South Korean economy, one focused on rapid expansion with its booming steel, shipbuilding, petrochemical, auto, and electronics industries, as it averaged nearly 8 percent annual growth. Oil demand rose by only 2.7 percent annually from 1980 to 1987, but averaged about 20 percent annually from 1987 to 1995, though economic growth was about the same (7.8 percent and 7.9 percent, respectively in both periods).[41] No less intriguing is the fact that oil decreased as a share of South Korea's total energy mix in the first period, yet comprised 85 percent of the increase in total energy demand in the latter period.

One persuasive analysis suggests several factors were responsible. The oil shock of 1979–80 and resulting price hikes dampening demand along with government efforts to foster more energy efficiency and conservation were a large factor in the slower growth phase.[42] A military coup that ousted Gen. Park Chung Hee in 1980, with more ambitious development plans, may also have played some role. In the latter period, the collapse of oil prices in the mid-1980s was one factor. This occurred amid Korean economic growth hitting high gear, and as an urban middle class was mushrooming. This meant a surge in electricity demand, which meant more reliance on oil-fired power plants. But its exploding oil demand was in no small measure the result of both burgeoning internal transport needs and widespread diffusion of private automobiles as incomes more than doubled. During this period, car ownership increased by 25 percent annually. From 1987 to 1995, oil demand increased by 20 percent annually.[43] The Korean case is all the more remarkable because by the mid-1980s it was already a relatively urbanized, industrializing economy, yet its oil consumption quadrupled in the decade before the economic crisis.

Oil Strategies

While growth in oil consumption is not expected to continue at anywhere near the 1975–95 pace, oil demand is projected to increase by up to one-third by 2010, roughly an additional 600,000 b/pd.[44] Despite its best efforts, South Korea remains heavily dependent on the Gulf/Middle East, which supplies about 75 percent of its oil. Seoul has sought to increase imports from Southeast Asia, Africa, and Latin America, but these remain small portions of South Korea's total energy picture and are unlikely to displace significant amounts of Middle East crude over the coming decade.

Beyond seeking alternative oil suppliers, Seoul has pursued a two-pronged energy security in regard to oil, again, one strikingly similar to that of Japan. It is gradually building up its strategic petroleum reserve, perhaps the most reliable assurance against short-term supply disruptions. Its current stockpile is the equivalent of a 60–day oil supply; the Korean government is constructing additional storage facilities and plans to increase the stockpiles to 90 days' worth of oil, which would meet the OECD requirement.[45] The other prong is the effort to obtain equity stakes in foreign oil and gas exploration and production to enhance its sense of energy security. The Korean National Oil Corporation has 18 exploration and production projects in 12 countries overseas, 4 producing fields, and is pursuing exploration projects in Vietnam and Indonesia. But like Japan's government-subsidized efforts, it has only resulted in modest achievements, less than 50,000 b/pd.

Even before the opening of Soviet–South Korean relations in 1990, Korean firms demonstrated a strong interest in developing oil and gas in the Russian Far East. Hyundai chairman Chung Ju Yung's visit to Moscow in 1989 led to an "agreement in principle" to develop resources in Yakutia. This evolved after the breakup of the USSR into a broader pledge of cooperation, but still not much more than a statement of intent. This was embodied in a Russian-Korean statement during a November 1992 visit to Seoul by then Russian president Boris Yeltsin. Yeltsin and then Korean president Kim Young Sam pledged to pursue feasibility studies for joint development of natural gas in Yakutia, the construction of pipelines to transfer the gas to Korea, and similar efforts to develop offshore Sakhalin natural gas. Nine Korean firms formed a consortium to begin the process. But difficulties in reaching agreement on feasibility studies, part of a larger problem with doing business in Russia, has led to naught thus far.[46] In the 2010–2020 timeframe, if sufficient quantities are available, Russian gas could conceivably be piped into Northeast Asia. Several pipeline routes have been suggested. One from Yakutsk through Vladivostok, North and South Korea, and then on to Japan via Pusan. Another would go from eastern Siberia through Mongolia, to Beijing. But the limited size of the Korean market will require cooperative ventures, most probably financed largely by Seoul and Tokyo and/or Beijing to reach an economy of scale for any of the ambitious Northeast Asian pipeline schemes to make economic sense.

Natural Gas Future

Natural gas will loom larger in South Korea's energy future, though in the near term, principally in the form of LNG. Since 1987, South Korean natural gas demand has increased by roughly 20 percent annually, accounting for 10 percent of world LNG consumption by 1999. Unlike the case with oil, the bulk of Korean LNG has been supplied by East Asian sources, with Indonesia and Malaysia accounting for more than three-fourths of its supply. However, recent long-term (25-year) contracts with Qatar and Oman will diversify its gas sources, with Middle East sources accounting for more than 40 percent by 2001, and the Indonesian and Malaysian shares would be reduced to under 50 percent, with Brunei and Australia providing small amounts.[47]

South Korea is substantially better positioned than Japan to rapidly expand the use of natural gas. It has invested heavily in gas distribution trunklines that comprise a distribution network in several major cities including Seoul and Inchon. It already has two receiving terminals, with plans to expand the capacity of both. In addition, Seoul is in the planning stages of a third receiving terminal and Enron and SK Corporation are expanding the gas infrastructure to five additional Korean cities.[48] This diverse use of gas is reflected in the breakdown of Korean demand: unlike Japan, where gas is largely used for electric power generation, in Korea, residential use and industrial use comprise more than 50 percent of Korean demand. Forecasts for Korean demand growth range from 5 percent to 11 percent annually to 2010.⁴⁹ Apart from the pace of economic growth, another important factor affecting gas demand growth will be the degree to which the Korean gas and electricity industry is restructured. Seoul announced plans for privatizing public companies, including the gas monopoly, KOGAS and KEPCO, the largest utility, and others. In theory, this will create an open access system, separating out importing from sales of gas and electricity, and opening the door to IPPs by 2002–2003.⁵⁰

The Importance of Nuclear Power

One of the most important pillars of Korean efforts to diversify supply sources, more even than natural gas, has been nuclear energy. Since its first commercial power reactor was commissioned in 1978, spurred in part by the 1973–74 oil crisis, nuclear energy use has grown dramatically. South Korea's 14 nuclear power reactors at four nuclear power plants generate 44 gigawatts of electric power.⁵¹ Nuclear energy accounts for roughly 40 percent of Korean electricity; 12 percent of South Korea's primary energy consumption. Seoul plans to have 16 reactors operating by 2010, but both the pace and scope of future nuclear power plant construction is problematic.

While less beleaguered by public ambivalence than Japan, Korean nuclear expansion plans do have to contend with the public acceptance question. There has been persistent public concern over nuclear safety and the problem of nuclear waste. By an odd coincidence, shortly after the Tokaimura accident, a Korean nuclear plant also had an event where radioactive water leaked from one its reactors.⁵² The first such accident at a Korean nuclear plant, it generated a wave of protest from environmental groups.

And as with Japan, nuclear power inevitably intersects with security questions. In the case of South Korea, this has a peculiar dimension to it because of the North Korea problem. Even before North Korea's nuclear weapons program became a serious regional concern, South Korea had covertly sought to pursue a nuclear weapons option in the mid-1970s. This was quashed under considerable U.S. pressure. Yet Korea's scientific intelligentsia retains an interest in the fuel cycle.⁵³ This is apparent in numerous discussions the author has had with both national security and nuclear energy analysts and planners during the 1990s. Though Seoul signed a 1992 denuclearization accord with North Korea wherein both sides agreed to forgo reprocessing, if Japan realizes its much delayed fuel cycle plans, this would both legitimize it in the eyes of some of the Korean elite and stimulate interest in pursuing a similar course.

The North Korea Factor

Nuclear concern is, of course, a major issue in regard to the real wild card in Korea's energy equation, North Korea. This was evidenced in the 1994 nuclear crisis, which led to a deal between Washington and Pyongyang to freeze and eventually dismantle North Korea's nuclear weapons program in exchange for a number of largely energy-related blandishments. But the nuclear issue that North Korea raises is but one aspect of the larger question of national reunification. The paradox here is that while Korea is a divided nation, much like Germany or Yemen were, and will inevitably be reunified, it is impossible to predict either how or when reunification will occur. After a decade of negative growth that saw its economy shrink by nearly two-thirds, North Korea is a failing state with enormous problems ranging from widespread hunger to massive energy shortages. Yet it is conceivable that Pyongyang could stagger on for another decade or it could collapse tomorrow—and it is impossible to predict either course.

The preferred course for reunification is a gradual process of détente, peaceful coexistence and increased economic interchange. But the communist regime in Pyongyang has been reluctant to more than experiment at the margins with economic reform or opening. The premise of a gradual process of reunification is that of a "soft landing." That hope is what prompted South Korean president Kim Dae Jung to call for a June 2000 summit with North Korean leader Kim Jong Il. At the other end of the spectrum of possibilities is the "hard landing," a sudden collapse of regime or state resulting in absorption by the South. The latter is a scenario Seoul desperately hopes to avoid. Indeed, all the major players in Northeast Asia, the United States, China, Japan, as well as both Koreas, have been pursuing what might be best dubbed a "muddle through" approach. Fearing the likelihood of conflict and/or chaos in the event of a collapse in North Korea the key actors in Northeast Asia have been placing North Korea on a life support system: China supplies food, coal, and oil; the United States supplies food and heavy oil for heating; South Korea also provides a range of economic benefits.

Energy has been a central element in North Korea's decay. It suffers chronic electricity shortages, constant fuel shortages, and highly inadequate heating. Its energy grid is crumbling and highly inefficient. Of almost 500 electricity generation facilities, only about 60 are part of a transmission and distribution grid. Yet North Korea insisted on two LWRs as the price for freezing its nuclear weapons program, rather than energy assistance that might have garnered more near-term payoff such as coal-fired thermal plants. In an "Agreed Framework" that was signed in 1994, the United States agreed to provide North Korea with 500,000 tons annually of heavy fuel oil until its LWRs are completed, initially planned for 2003, but an unexpectedly complex process combined with intermittent mini-crises have set back completion to no earlier than 2007–2008. Even then, the reactors have no connecting energy grid to distribute power. Indeed, the reactor project only makes sense in a scenario of North-South economic integration, where they would be part of an integrated nuclear power program. but no such reality exists or is on the horizon. In the meantime, North Korea is stockpiling the majority of the U.S.-supplied heavy fuel oil that it cannot absorb. It relies on oil for only 8 percent of its energy needs, the bulk of its energy produced by coal.[54]

In any case, reunification would likely mean a wholesale reconstruction of North Korea's economy. Cost estimates range from $500 billion to $2 trillion.[55] This depends on what set of assumptions the respective analyst employs—how rapid, how thoroughgoing reconstruction is assumed to be. But in regard to energy, the point of this discussion of North Korea is twofold: that reunification would dramatically change South Korea's energy equation; and that while it looms as a future likelihood, it is a problem deferred at present. Suffice it to say that reunification would be an almost unfathomable burden on the South Korean economy. At the same time, it would almost certainly ramp up energy demand to levels similar to those Korea saw during the 1980s, after an initial period of uncertainty, as what was North Korea's energy infrastructure would have to be reconstructed from the bottom up, with Seoul having new responsibilities for the North's entire energy needs.

Conclusion

While there is keen awareness of energy vulnerabilities in South Korea, as in Japan, beyond the energy policies discussed above, there is little evidence of any broader planning to address energy insecurities. Indeed, there appears an implicit sense that should worst-case scenarios unfold that threaten the flow of energy resources, the South Korean military would have little ability to respond. A report prepared for the Pentagon's Office of Net Assessment concluded, "There is a strong sense in the [Korean] military that the United States will always be there to solve a Korean energy crisis. . . . South Korea's role in any energy security scheme can only be in support of a larger multilateral effort." [56] But South Korea's energy security policies may also be based on an implicit assumption that old notions of energy security may no longer apply in a twenty-first-century Information Age global economy. The approach discussed above is more consistent with the latter energy security calculus than the former.

Chapter 8

Southeast Asia: Tigers as Regional Energy Link

The enormous, spiral-like Petronas Towers extend nearly half a kilometer up into the sky. The world's tallest buildings when completed in 1998, the towers dominate the Kuala Lumpur skyline. A symbol, if a tad garish, of pride and presence for Malaysian president Mahatir Mohammed, they all but scream out: "Look at me, I'm important." The buildings may be something of a metaphor for Southeast Asia's 's desire to establish a higher profile. It is difficult to think of 500 million people with a collective gross national product of some $700 billion as an afterthought. But placed in the larger context of the Asia-Pacific, the ten developing (with the exception of Singapore) states in the Association of Southeast Asian Nations (ASEAN) do tend to be overshadowed by the sheer size of China and India, the tense security drama of the Korean peninsula, and the economic heft of the world's second largest economy, Japan.[1] The gleaming skyscrapers of Petronas, Malaysia's ubiquitous state oil conglomerate, are emblematic not only of Southeast Asian aspirations but the region's energy policy challenges as well.

Yet Southeast Asia, while a relatively minor energy consumer, by virtue of geology and geography plays a pivotal, multifaceted role in the East Asian energy security equation. The story of energy in Southeast Asia is principally one of oil and gas. The subregion includes significant oil exporters (Indonesia, Malaysia, Vietnam) and major intra-Asian natural gas exporters (principally Indonesia, Malaysia, and Brunei exporting to Northeast Asia). Singapore is a major global and regional oil and financial entrepôt, the world's third largest refining center as well as a trading and price-setting nerve center. Indonesia, the linchpin of subregional energy flows, sits astride the world's major shipping lanes (and strategic chokepoints), the straits of Malacca, Sunda, and Lombok, through which some 250 ships a day pass with oil from the Persian Gulf and an array of other goods worth roughly a trillion dollars each year and amounting to some 40 percent of world trade.[2] These straits are surrounded by the South China Sea, through which these ships also pass en route to China and Northeast Asia and about which territorial disputes remain a source of potential conflict and an obstacle to developing potential oil and gas resources. (See Figure 8.1.)

Prior to July 1997, when a run on the Thai Baht triggered what quickly became a severe Asian economic crisis (discussed in chapter 4), many of the ASEAN economies—Thailand, Indonesia, Malaysia, and possibly the Philippines and Vietnam—appeared poised to become the third wave of the "Asian miracle." Following Japan's economic success into the 1980s, Singapore was already one of the "Four Tigers" (along with South Korea, Hong Kong, and Taiwan), whose sustained, rapid export-led growth forced

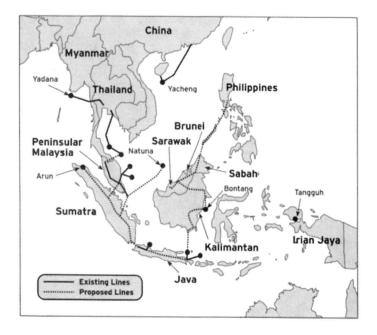

Figure 8.1 Existing and Proposed Pipelines in Southeast Asia
Source: James T. Jenson, "Natural Gas Policy Issues for the Asian Region," in Dona K. Lehr (ed.) "Natural Gas in Asia: Facts and Fiction," Published by Pacific Economic Cooperation Council, 1998.

many to question assumptions about models of growth for developing countries. In the late 1980s and early 1990s, Thailand was one of the fastest-growing economies in the world, expanding by more than 10 percent annually. Broadly, these core ASEAN economies (Indonesia, Malaysia, Singapore, Thailand) have averaged over 6 percent annual GDP growth from 1975 to 1997.[3]

But in the aftermath of the 1997–98 economic crisis left a large measure of uncertainty in its wake and complicated any projections about future growth. All the Southeast Asian economies did display unexpectedly buoyant recoveries in 1999. In the case of Thailand and Indonesia they rebounded from −9.4 percent and −13.7 percent contractions, respectively, in 1998 to roughly 5 percent growth in the case of Thailand and near 0 percent for Indonesia. But these recoveries must be considered tentative. In a perverse sense, Thai and Indonesian short-term recovery may render sustained long-term growth more problematic, as such developments reduce the pressure for structural reform that would address the economic fault lines revealed by the 1997 crisis. Thailand and Indonesia, for example, have yet to reach agreements with creditors to write down their respective debt, which comprises nearly half of Thai bank lending and in the case of Indonesia totals 110 percent of GDP. Moreover, apart from their own domestic predicaments, as oil is purchased in dollars, the lingering currency weaknesses in much of Southeast Asia mean that oil prices, which for most of 1999 and to mid-2000 have been in the $25 to $30 a barrel range, pose significant challenges to Asian recoveries—though oil price hikes boosted the fortunes of exporters like Indonesia, oil prices compounded economic problems for countries such as Thailand and the Philippines. This analysis accepts the International Energy Agency (IEA) projection of 4.1 percent overall primary energy demand growth in the 1995–2020 period, with oil averaging 3.6 percent annual growth and gas 5.5 percent.

Southeast Asian Energy

To put ASEAN's place in the global fossil fuel picture into perspective, the subregion has just under 2 percent of world oil reserves and accounts for 3.5 percent of world oil production.[4] (See Figure 8.2.) It has just 4 percent of proven world gas reserves, some 3,600 billion cubic meters, but accounts for 20 percent of world production.[5] Though ASEAN states are only minor oil suppliers to the major Asian consumers in Northeast Asia, they are key suppliers of liquefied natural gas (LNG) to Japan, South Korea and Taiwan, who account for 75 percent of world LNG consumption. Indonesia alone supplies 40 percent of Japan's LNG and is also a modest but growing exporter of high-quality, low-sulfur coal to Asia-Pacific nations.

On the receiving end, ASEAN states accounted for just over 3 million b/pd of oil in 1999, about 15 percent of the East Asia's total consumption. ASEAN also accounted for roughly 70 billion cu. m. of natural gas, over 20 percent of East Asia's gas use, a proportion of the region's consumption that will diminish sharply over the next two decades as India and Northeast Asia expand the role of gas in their respective energy mixes. While the reserves-to-production (R/P) ratio for gas is a healthy 41 years, ASEAN oil producers have unusually low reserve-to-production ratios, averaging less than 12 years (the world average is 41 years). This has led many forecasters to predict that Indonesia and other Southeast Asian producers will become net oil importers in the course of the coming decade.[6] Of all the Asia-Pacific oil producers, only the Southeast Asians are still net exporters. Both China and Australia have increased their production over the past two decades, and their respective consumption has grown still faster, though Australia is a net gas exporter, with more than one-third of its gas sold abroad.

As discussed in chapter 4, reports of the impending decline of Asian oil production may be significantly exaggerated. The key variable will be the degree to which Asian governments create more hospitable investment environments in their oil sectors. Nationalist approaches to managing oil, expressed in the form of state monopolies such as PetroVietnam and Pertamina in Indonesia controlling all upstream and downstream oil activities, have created impediments to foreign investment and the application of new technologies to prolong the life and expand the capacity of existing production. While there appear to be geological actuarial tables facing ASEAN producers, this is to a considerable extent a matter of policy and politics.

1997 Oil Production, Oil Reserves, and Gas Production of Selected Southeast Asian Nations

Country	Oil Production (thousand barrels daily)	Proven oil Reserves (thousand million barrels)	Gas Production (billion cubic feet)
Brunei	175	1.4	391
Indonesia	1,602	5	3,132
Malaysia	751	3.9	1,376
Papua New Guinea	120	0.3	4
Thailand	123	0	571
Vietnam	191	0.6	18

Figure 8.2. 1997 Oil Production, Oil Reserves, and Gas Production of Selected Southeast Asian Nations
Source: International Energy Agency, U.S. Energy Information Administration, BP Statistical Review, 1999.

In this regard, it is worth underscoring again the contrast between the fate of North Sea oil over a recent two-decade period (1975–95) and that of Asian oil. In 1975, both had similar proved reserves, both of which were significantly underestimated. Yet in the intervening period, particularly by the 1990s, both North Sea production and reserves were substantially greater than Asian production, though the performance of the North Sea fields was not the result of new discoveries. Instead, a careful, technical analysis reveals the differences to be principally the result of "cost saving for any given state of technology, advances in technology and changes in the fiscal regime."[7] Investments in applying advanced drilling techniques, horizontal drilling, and the development of 3–D seismic and 4–D seismic have all made major contributions to the North Sea's prediction-defying longevity. Paul Hornswell, in a compelling assessment, concludes that "the two most important determinants of the future path of Asian production are fiscal conditions and the industrial organization of the upstream sector . . . the relatively poor performance of Indonesia is very much a function of its extremely hostile fiscal regime." He adds that "there would be the potential of very significant gains in production in Asia . . . if there were a liberalization of the Asian upstream."[8]

Southeast Asian producers have pursued government policies that, for example, allocate prime fields to state-run monopolies, relegating foreign investment to marginal fields. Such practices combined with convoluted and arcane fiscal and pricing policies that have in some cases reduced the netback (profit minus production, transport, and tariff costs) such that foreign investors have withdrawn and potential investors have been kept away. As one prominent Western oil executive active in the region put it, "One issue for the future are the fiscal terms offered by host governments which are harsh by world standards, with limited incentives for oil companies to explore, particularly in frontier areas. Are improvements to the various tax regimes the key to help delay the inevitable decline in oil production?"[9] It must be added that these impediments to developing oil resources are multiplied several-fold in regard to natural gas. The substantial capital costs in developing natural gas—whether LNG or pipeline gas—the somewhat arcane process of pricing gas, the scale of economies, and the long-term (20–30 years) commitments required for markets, make it substantially more complex to develop gas fields.

In the interim, the result has been the opportunity costs of not obtaining the capital and technology to maximize production and expand reserves in existing fields by fully applying the remarkable Information Age technologies that have revolutionized the oil and gas business over the past two decades. While the volume would be marginal to the larger Asia-Pacific energy equation, the higher output would make a significant difference in the energy calculus of producer countries. Few in the region have paid higher opportunity costs than Indonesia and Vietnam in this regard.

Indonesia and Singapore: Upstream and Downstream Linchpins

Indonesia, the sprawling 13,000 island archipelago that is the world's fourth largest nation, is the keystone of Southeast Asia not just as the leading oil and gas player but as the world's largest Moslem nation, the only Asian member of OPEC, and the dominant geopolitical actor and the bulwark of ASEAN as a regional political and economic institution. Yet in 1999–2000 it was also the region's largest question mark. The twin economic and political crises of 1997–98 ended in the demise of three decades of President Suharto's authoritarian rule and a deep economic recession, with 75 percent of Indonesian businesses technically bankrupt as a result of the currency crisis and a $43 billion International Monetary Fund emergency debt-relief package to keep its wounded economy afloat. While its currency, the rupiah seemed to stabilize at 7,000–7,500 to the dollar—from 2,500 before the crisis—in mid-2000, it was at a level that still greatly complicates efforts to renegotiate its corporate debt. Moreover, Indonesia has been plagued by sporadic violence and political unrest in several provinces as a fledgling democratic government seeks to virtually reinvent an entire political culture frozen in an authoritarian, military-dominated mode for thirty-three years.

Indonesia's turmoil has had a variety of ramifications for its energy sector. Before a democratic election was held in October 1999, the caretaker regime of B. J. Habibie acceded to a popular referendum in East Timor, though the process was marred by savage outbreaks of violence by local militias backed by provincial military forces. The former Portuguese colony, which the Indonesian military occupied and annexed in 1975 (fearing a prospective handover to a Marxist guerrilla group upon Lisbon's withdrawal), was then severed from Indonesia. East Timor was administered by a UN-led administration and multinational peacekeeping force during its transition to independence. Among the consequences of East Timor's independence is uncertainty about the fate of the Timor Gap Treaty, under which Indonesia and Australia negotiated agreements to divide revenues from oil and gas in the area. It is not clear whether the accord will be renegotiated with East Timor, or if the independent government in Dili will become the successor state to the treaty. Thus far, only modest amounts of oil (33,000 b/pd) have been found, this in a Zone of Cooperation where revenues are to be split 50–50. There have been a number of other modest discoveries as well as indications of significant amounts of gas, some of which are in early stages of development.[10]

The large pockets of unrest across Indonesia raised the fear that the nation could unravel in an Asian version of the breakup of Yugoslavia. Most prominently, violence and demands for independence in Aceh, a major gas-producing province at the northwestern tip of Sumatra posed a difficult challenge to Indonesian President Aburrahman Wahid. Unlike East Timor, Aceh has been an integral part of Indonesia since it gained independence from Holland in 1949. Virtually every oil and/or gas rich province of Indonesia—West Irian, Riau, and East Kalimantan—raised demands ranging from more autonomy and larger shares of oil and gas revenues to independence, underscoring the centrifugal forces that Jakarta is seeking to reassert authority over. By mid-2000, however, President Wahid appeared to be slowly consolidating his political control and quietly, if gradually, ameliorating the worst-case fears about Indonesia's future by offering new social bargains to key provinces and removing the Indonesian military from its dominant role in the nation's political life.

The political turbulence and economic uncertainty in Indonesia make it precarious to assess the direction of its energy policies. Prior to the 1997–98 crisis, Indonesia had averaged 6.3 percent annual economic growth for two decades. It was one of the few oil-rich states that manage to diversify its economy: by the 1990s, oil and gas accounted for less than 30 percent of Indonesian exports. Projections (based on sustained growth levels) made prior to the crisis were that Indonesian demand for oil and gas would triple between 1995 and 2015, with Indonesia becoming a net energy importer by 2003–2005. Yet in the decade 1984–94, Indonesian oil consumption grew by only 71 percent.[11] In any case, Indonesia economic growth in the 2001–2004 time frame will almost certainly be well below pre-crisis levels, thus easing oil demand.

Indonesia's hydrocarbon resources (oil and gas account for nearly 90 percent of domestic demand) will be adequate to meet domestic needs over the next two decades under even high-growth scenarios. But Indonesia's fate as an oil exporter is less certain. Its production has been stable at about 1.5 mil b/pd for more than a decade, and in 1999 Indonesia's oil consumption was 930,000 b/pd. It sits on 5 billion barrels of proven oil reserves, but they have declined 14 percent since 1994, and Indonesia's R/P ratio is only 9.2 years.[12] As discussed above, this situation is not necessarily permanent but depends on the amount of investment that goes into the oil sector. That in turn depends largely on Jakarta's ability to make the necessary policy adjustments to lure more foreign investment (e.g., tax breaks, more favorable production-sharing agreements).

Indonesia has bountiful natural gas reserves, 72.3 trillion cu. ft. (Tcf), and substantial finds in the late 1990s—such as ARCO's West Irian Tangguh fields, worth more than 13 Tcf of gas reserves—suggest still larger possible reserves will be developed. In 1999, for example more than 3 Tcf in gas was discovered on Sulawesi. Already, Indonesia is a pivotal supplier of LNG to Japan, South Korea, and Taiwan. Asia's burgeoning use of natural gas in the first decade of the twenty-first century will not only be an important catalyst in rehabilitating the Indonesian economy but also will provide Indonesia with a low-cost domestic alternative to oil as a source of energy. One of the most significant finds in recent years is the D-Alpha field (the largest in Southeast Asia), offshore Natuna Island, being developed in a $40 billion project. However, the gas is very impure, with a 70 percent CO_2 content that will complicate its development. Indonesia has signed contracts to export gas from West Natuna to Singapore via a 300–mile pipeline, construction of which began in 1999.[13] Unlike the case with oil, which threatens to be depleted in the foreseeable future, Indonesia has 30 years R/P ratio in gas, one that is likely to grow significantly if, as expected, gas continues to be discovered.[14]

The Pertamina Factor: Post-Suharto Energy Policy

Much of the fate of Indonesia's energy industry and its energy policies hinges on the fate of Pertamina, the state-owned oil and gas monopoly. Pertamina is a prototypical state monopoly not unlike a number of other developing country monopolies. It has controlled all upstream and downstream activities in Indonesia's petroleum, operating somewhat like a state within a state. It thrived off disproportionate production-sharing agreements that foreign investors in the energy sector were required to reach with Pertamina. Production-sharing agreements with more than 70 foreign energy firms have allowed Pertamina to garner 85 percent of the revenues.[15] Not long after its founding in 1969, Pertamina became a cash cow, an important tool for dispensing patronage to family and cronies under Suharto. Over the course of Suharto's tenure, Pertamina awarded 159 contracts to companies linked to his family and cronies, all without formal bidding or negotiations. Pertamina's pressure leading operators to be obliged to select approved subcontractors reportedly added some 20–30 percent to the cost of oil and gas projects.[16] So inefficient and scandal ridden was Pertamina that it was pressed to launch a restructuring program in 1993 that reduced production costs to $5.1 a barrel from $11.7, and reduced its work force to 30,000 from 45,540. In July 1999 an independent audit of Pertamina by Price Waterhouse Cooper revealed that graft and inefficiency cost the state firm $6.1 billion during 1997–98.[17] In the first six months after Suharto's downfall in May 1998, the company saved $100 million by tearing up questionable supply and distribution contract with 32 firms, many of which were linked to the Suharto crony network.[18]

Efforts to dismantle or radically reform Pertamina were begun in 1999 when legislation aimed at bringing Indonesia's oil and gas industry up to international standards was introduced in its parliament. The reform legislation sought to remove Pertamina from overseeing production-sharing contracts and deregulate both upstream and downstream sectors, essentially removing Pertamina's middle-man role between producers and consumers. The legislation would move the awarding and supervision of production-sharing contracts to the Minister of Mines and Energy, and also free foreign oil firms from a welter of regulatory constraints.

In mid-2000, the precise final contours of reform of the oil and gas industry were still unclear. There is no question, however, that the petroleum industry is moving in a more market-oriented direction. In January 2000, President Wahid appointed a new commission to oversee the restructuring of Pertamina. The Wahid administration has stated its intention to pursue legislation along the lines of that introduced prior to Wahid's accession to office, a process likely to play out in the 2000–2001 time frame. These reforms will have a significant impact on Indonesia's energy strategy, though political considerations render a number of issues problematic in the near term. The downstream reforms will be somewhat arduous and likely phased in over time. The fragile state of the Indonesian economy makes it extremely difficult to end fuel subsidies, and with 70 percent of its workers in refining and distribution sector, reform there will also likely be gradual. Another sensitive political issue is the distribution of revenues from production-sharing agreements. Wahid has promised larger shares of oil and gas revenues will be shared with oil- and gas- producing provinces. But this will be a delicate balancing act, as parceling out substantially more revenues to the provinces than in the past would drain Jakarta's already thin treasury as it seeks to meet IMF targets.

In any case, reform will facilitate the implementation of a post-Suharto energy policy that follows three broad themes: expand supplies by increasing productivity of energy resources, diversify energy sources (particularly the oil/gas balance), and encourage conservation through more energy efficiency. One major shortcoming of energy policy under Suharto was a lack of emphasis on gas rather than coal and oil for electricity generation. Gas is still flared off in fields within sight of power plants. Oil still supplies roughly 50 percent of Indonesia's primary energy needs. Of the gas it uses, about 50 percent is used in its industrial sector, only 37 percent for power generation.[19] But expanding the use of gas requires large capital investments in infrastructure for transmission and distribution, particularly for an island/archipelagic nation. Prior to the economic crisis, Jakarta had ambitious plans to build a national gas grid including a trunk gas pipeline to link Java and Sumatra, which together account for more than 75 percent of the country's population. But such plans are not realizable in the near future in light of Indonesia's debt and troubled economy. However, in a 2005–2010 time frame, transition to more gas turbine combine cycle power generation promises dramatically increased energy efficiency, lower costs, and hard currency earnings by freeing up oil for export. In the near term one of Indonesia's larger problems in the electricity sector is the loss of confidence generated by efforts by PLN, the state-owned electricity utility to renege on contracts with foreign suppliers in the face of massive debt due to collapse of the rupiah.

Indonesia's energy agenda, however, looks beyond hydrocarbon resources to renewable energy sources for its energy security. Of Indonesia's 20 gigawatt (GW) electricity generating capacity, 15 percent is from hydropower and 3 percent from geothermal sources. Indonesia completed its largest hydroelectric plant in April 1998, a 1,000 MW plant in West Java, but with its economic downturn, Indonesia has neither the finance nor the demand for further hydropower expansion in the near future. Moreover, Indonesia boasts the world's largest potential for developing geothermal energy, some 20,000 GW, though it currently utilizes only 400 MW. Prior to the economic crisis, Pertamina had approved 11 geothermal projects with probable support for some from Japan, but these projects are also slowed, if not dormant, in the economic environment of 2000. In addition, over the longer term, Jakarta had actively considered a nuclear energy option. Indonesia already has three research reactors and in the mid-1990s was contemplating building up to 12 nuclear power plants in Java. Such ideas are now dormant and unlikely to be revisited before the 2020–2030 timeframe, if at all.

Singapore, Petroleum Central

If Indonesia is the region's resource linchpin, Singapore is its commercial/processing linchpin. Its strategic location and excellent deepwater harbors have given Singapore a comparative advantage as an entrepôt in the world trading system since the British first arrived more than 150 years ago. Its $100 billion annual GDP economy generated by only 3 million Singaporeans who have reached OECD level incomes (average per capita GDP: $25,000) is an important example of relative unimportance of territory for economic prosperity in the global Information Age, a point worth underscoring in light of the myriad territorial disputes in East Asia that threaten regional stability. Utilization of comparative advantage, well-conceived economic management, judicious use of industrial policy, and policies fostering foreign investment and export-led growth have been key to Singapore's success.

The petroleum industry, 11 percent of its manufacturing sector, has long been part of the Singapore story reaching back to the early days of the oil era. In the 1870s, Singapore became a regional distribution center for tins of kerosene lighting up the lamps of the East. A generation later, it became a Far East operations center and storage facility initially for Royal Dutch and Shell companies.[20] Today, petroleum is part of the reason why the Port of Singapore is the world's largest in terms of shipping tonnage. The rise of Singapore energy sector parallels its emergence as part of the "Asian Miracle." Since 1960, Singapore has become the third largest refining center in the world, with a capacity of 1.2 million b/pd. Its free-market philosophy when much of Southeast Asia was immersed in postcolonial nationalism helped make it a magnet for overseas investment.

Singapore's wherewithal as a regional commercial and financial hub as well as its refining capacity and reliability has facilitated its evolution first into a third-country processing center, and in the 1990s its focus on niche oil products (middle distillates) helps explain its buoyancy in the petroleum business.[21] More recently it has utilized its refinery base to expand into a massive petrochemical complex. A government project to triple the landmass on reclaimed land at Jurong Island, cementing together adjacent tiny islets connecting by a four-lane causeway to the main Singapore island, promises to result in a major regional petrochemical hub by 2003–2004 time frame. Already there are some $9 billion in major investments by U.S., European, and Japanese multinationals in the complex.[22] Clearly, Singapore has bet its energy security on global markets.

Modest Producers: Malaysia, Vietnam, Brunei

It may not quite measure up to the pretense of its Petronas Towers, but Malaysia is also an important and multifaceted player in the Southeast Asian energy game. Its 3.9 million barrels of proven oil reserves and 81.7 Tcf of natural gas are surpassed only by Indonesia among ASEAN countries. Like Indonesia it has a rather low R/P ratio of oil reserves of 25 years, which similarly raises the question of whether Malaysia may become a net oil importer by roughly 2010.[23] But Malaysian natural gas has a relatively healthy 30-year R/P ratio.

Unlike Indonesia, however, Malaysia has steadily increased its oil production—from 703,000 b/pd in 1995 to just under 800,000 b/pd in 1999.[24] Though a victim of the Asian economic crisis that disrupted 13 years of economic growth averaging 7.5 percent with a -6.7 percent contraction in 1998, Malaysian oil consumption has continued to increase, reaching 450,000 b/pd in 1999 (up from 378,000 b/pd in 1994).[25] Moreover, Malaysia consumed some 40 percent of the 1.45 Tcf it produced in 1999, largely for power generation. Malaysia is on an economic growth trajectory that may see it reach OECD GDP levels by 2020. This will mean substantially increased per capita consumption levels of oil and gas over the coming decade and the one thereafter that will result in a steady reduction in, and perhaps culmination of, Malaysia's status as an exporter of petroleum products within this time frame.

Malaysia is a particularly interesting energy actor in several other respects. It offers a potentially important precedent in its energy cooperation with Thailand. One of the areas of most active gas exploration is the Malaysia-Thailand Joint Development Area (JDA) along both countries' continental shelves in the Gulf of Thailand, which so far includes 10 Tcf in estimate reserves. The example of bilateral cooperation in energy development in once disputed territory is one with potentially larger implications for energy security in light of ambitious ASEAN plans for a Free Trade Area by 2003 and an ASEAN energy grid (see below). Moreover, a $1.03 billion infrastructure project including a binational pipeline network and a natural gas processing complex is further concrete evidence of operational energy cooperation. For all the speculation in Asia about pipeline networks, the Trans-Thai-Malaysia project is a rare instance of common action to construct one. The first phase of the project is a 352 km pipeline from the offshore JDA to southern Thailand, due for completion before the end of 2001.[26]

The second distinctive feature of Malaysia as an energy actor is the phenomenon of Petronas, the state-owned oil and gas company. In sharp contrast to Indonesia's state monopoly, over the years, by most accounts, Petronas garnered a reputation for solid commercial management and profitability. It has also invested in virtually every facet of the oil industry.

Moreover, to compensate for diminishing reserves, it has invested some $8 billion in oil and gas equities in an array of more than 20 countries from Algeria to Vietnam. Petronas now earns more than one-third of its revenues overseas. Within the ASEAN region, apart from its investment in the Malay-Thai JDA, it also has a 30 percent ($200 million) stake in the offshore Burma Yetagun gas project expected to be piping gas into Thailand by 2001.

Yet even before the 1997–98 economic crisis, Petronas was occasionally used as an adjunct to the national treasury, bailing out failing companies. But recent politically directed ventures outside its core oil- and gas-related areas have begun to test its ability to sustain itself as a profitable, full-service energy company. In the aftermath of the crisis it helped buy debt-ridden shipping assets controlled by a son of the Malaysian prime minister. It has also accumulated a large portfolio as a property developer, and in 1999 geared up to buy controlling shares of Malaysia's national car, the Proton.[27] Though Petronas turned a respective profit in the midst of East Asian adversity, its numerous acquisitions turned it into a net debtor. Nonetheless, For a small nation of 17 million, Malaysia and Petronas have become significant energy actors, and in the process positioned themselves in regard to future energy security in the near to mid-term. Over the horizon, in the 2015–2025 time frame, Malaysia may face some significant energy policy challenges as it adjusts to the status of net importer.

In contrast, Brunei, a tiny sultanate of 300,000 people surrounded by the Malaysian province of Sarawak on the island of Borneo, is more like a mini-Asian version of Arabian Gulf mini-states like the UAE. It produces about 160,000 b/pd of oil, the bulk for export with a below-average R/P ratio of 24 years. It also has significant gas reserves, 13.8 Tcf, with a R/P ratio of 35 years. It produces 11 billion cu. mt. of gas, largely for export to Northeast Asia.

The other significant oil and gas actor in Southeast Asia is Vietnam, with 600 million in proven reserves and production that reached 250,000 b/pd in 1998.[28] With no refineries, it exports almost all of its production, largely to Pacific Rim nations, particularly Japan, South Korea, Singapore, and the United States. While it is believed to have larger potential reserves, its R/P ratio at present is only 6.6 years. The EIA suggests Vietnam may have the potential to produce 450,000 b/pd by 2020.[29] There have also been important gas discoveries in Vietnam in the mid-1990s, but while Vietnam is utilizing modest amounts domestically, it has been slow to develop gas production of significant volume.

Whatever Vietnam's oil and gas potential, it will likely be deferred until there are major policy changes in the direction of deregulation and more accountable governance. When Vietnam's Leninist leaders introduced initial reforms and opening in 1988, as in China, the economy responded, and the country grew an average of 8 percent annually from 1990–1997. While the economic crisis had some negative impact, the modest reforms allowing markets a larger role in agriculture and opening to foreign investment (much in the export sector) began to reach their limits and corruption mounted. Unlike the case in China, which has accelerated market reforms, the Vietnamese leadership backtracked. Not only did foreign direct investment begin to shrink dramatically—by 46 percent in 1999—but those already active in Vietnam began to withdraw. Vietnam's bureaucratic red tape, corruption, and lack of commercial law made business too difficult. In 1999, telecommunications firms Motorola, Qualcomm, and Cable and Wireless all left in frustration.

In the oil and gas industry, where the state monopoly, PetroVietnam, controls all upstream oil and gas ventures, companies also have begun to bail out. Shell withdrew in 1996. Vietnam's taxation system is inflexible and operates on a field-by-field basis, making its regulatory framework difficult for foreign oil companies. A number of western firms, including Exxon and Mobil, have withdrawn or suspended operations. In mid-2000 the most active foreign company was a Russian firm in a joint venture with PetroVietnam. For the near term at least, Vietnam's petroleum industry appears a reflection of its larger economic dilemmas and it remains in the "missed opportunity" category, or as a headline in the *Economist* read, "Goodnight, Vietnam."[30] However, after delaying in July 2000, it finalized a bilateral trade accord signed with the United States that will encourage reform.

Modest Consumers: Thailand and the Philippines

Thailand and the Philippines, two large Southeast Asian democracies that together have some 140 million people, collectively consume about 1.15 million barrels of oil per day, 5 percent of the total consumed by Asian nations in 1999. Thailand has nearly 300 million barrels of proven reserves and produces about 110,00 b/pd, but its requirements are nearly 800,000 b/pd. The Philippines imports about 98 percent of the roughly 350,000 b/pd of oil it requires, producing a tiny 4,000 b/pd, though recent offshore finds suggest some potential for increasing Filipino oil production.[31] But in the near future, it will not substantially reduce the million b/pd the two ASEAN states import, some 75 percent of which comes from the Persian Gulf.

Both possess significant amounts of natural gas, Thailand, 12.5 Tcf, and the Philippines, 2.9 Tcf. Ultimate recoverable gas reserves will almost certainly be somewhat larger; these quantities are inadequate for export but make an important contribution to both Thai and Filipino gas needs. Thailand's use of gas is far more developed, and Bangkok has placed priority on increasing the use of gas, 90 percent of which is used for electric power generation. Thailand produced about 525 billion cu. ft. (Bcf) of gas in 1999. With gas from the Malay-Thai JDA (discussed above) gas being developed in Burma's Yadana field, and arrangements made by the Petroleum Authority of Thailand (PTT) to import gas from Indonesia beginning in 2004, Bangkok appears to have excess gas supplies.

The Philippines also sees development of gas use as a priority, but with its 7,000 islands, constructing a national grid is a challenging task. Shell, which holds the concession at the Malanpaya-Camago offshore gas field near Palawan in the South China Sea, is building a 312–mile pipeline to link the offshore fields to three gas-fired electricity generation plants expected to come on line in 2001–2002. At 5.12 Tcf, the Malanpaya gas field represents the bulk of Philippine natural gas reserves.[32] The Philippines has depended on oil-fired plans for 47 percent of its electricity, geothermal (it is a world leader in this area) for 24 percent, and hydro for the remainder, but a number of its oil-fired plants are near the culmination of their service life. In the 2001–2005 period, unless Manila finds additional offshore gas, it is likely to require gas imports to meet its electricity needs.

The ASEAN Energy Security Future

Where does this energy picture leave ASEAN's energy security? As discussed above, scarcity does not appear an animating concern. Rather, there is a desire to increase energy security defined as diversifying supplies (particularly increasing intra-regional suppliers), increasing efficiency, and developing more environmentally friendly energy sources. There is a growing sense of shared interests reflected in a proliferating body of ASEAN coordinating mechanisms such as the ASEAN Council on Petroleum, the ASEAN Energy Minister's meetings, and the agreed objective of building an ASEAN power grid. ASEAN common interests include: a desire to reduce dependence on Middle East energy sources; to diversify their respective energy mixes in the direction of renewable sources; and to promote the use of gas both within ASEAN and within the Asia-Pacific Economic Cooperation (APEC) Forum. A "Vision Statement" issued at the 1997 ASEAN summit called for establishing: "interconnecting arrangements in the field of energy and utilities for electricity, natural gas and water within ASEAN through the ASEAN Power Grid and a Trans-ASEAN Gas Pipeline and Water Pipeline an promote cooperation in energy efficiency and conservation."

There is historical record of large gaps between ASEAN proclamations and reality. Such grandiose rhetoric, particularly in light of the post–1997 crisis economic constraints, might strike some as fanciful. As we have seen, there is no ASEAN grid nor an ASEAN pipeline, nor will there be such an energy network anytime in the near future. The enormous capital costs of lengthy pipelines require economies of scale.[33] This tends to produce a "chicken and egg" problem with regard to gas development.

The reality, however, is that piece-by-piece market-driven trans-border infrastructure and training networks are slowly becoming a reality. The Malay-Singapore pipeline, the Thai-Burma pipeline, the Indonesia-Singapore pipeline are all either operating or under construction. Close analysis of the APEC 1998 energy ministers meeting in Okinawa, which sought to advance an APEC "Natural Gas Initiative," reveals that APEC recognizes the paramount importance of the market in animating any such energy schemes. The recommendations focused on an array of legal, fiscal, regulatory, pricing, transport, and distribution policies necessary to generate private sector investment to create such energy cooperation.[34] At the risk of oversimplification, it may be said that these policies boil down to fostering competitive energy markets. To borrow from the movie *Field of Dreams*, this translates as, "If you build the right economic environment, investment will come."

While this gradual movement toward more self-contained regional energy development as an instrument of energy security unfolds, ASEAN is also living with several layers of vulnerability at the point where energy and security concerns intersect. Most involve not access to energy resources or assurance of supply; those are problems largely addressed by markets and policies (e.g., oil stockpiles). Rather, the vulnerabilities involve the physical control of supply. On one level there is the harassment of the problem of maritime oil piracy. There is an increasingly sophisticated hi-tech piracy carried out in the South China Sea area, with tankers being boarded and oil products transferred to other ships. Such brazen acts occur with some regularity—almost monthly, still more regularly for general shipping. While piracy does not threaten freedom of navigation writ large, there is a risk of environmental disaster as tankers and LNG carriers are targets of choice.[35]

Another ongoing interest is the risk of disruption of the shipping lanes, a low-probability, high-cost concern. A study of possible threats to the sea-lanes by the Center for Naval Analysis identified an array of threats starting with the nonmilitary: oil spills, bad weather, or volcanic activity disrupting one of the three critical chokepoints into the South China Sea (straits of Malacca, Lombok, and Sunda) or terrorists planting mines. There are also a host of military scenarios: naval or mine-laying blockade; Southeast Asian states impede the sea-lanes; military conflict over competing territorial claims; Chinese aggression over the territorial claims; China's blockade of Taiwan.[36]

None of the human-initiated scenarios were judged to be highly likely, though conflict in the South China Sea between China and either Vietnam or the Philippines, and between China and Taiwan, were higher on the risk scale. In the case of conflicts affecting the South China Sea passages, detours through the Philippines Sea, or in the worst-case around Australia would permit shipping to proceed, albeit at higher cost. While it cannot be denied that any of the above disruption scenarios is possible, it should be kept in perspective. Blocking the freedom of navigation in the World's major trading routes would be a grave act threatening the vital interests not only of ASEAN states but of all the Pacific Rim powers, not least the United States, Japan, and China. Indeed, such action could be viewed as the equivalent of an economic weapon of mass destruction, disrupting the functioning of the global economy. This raises the question of motive and underscores why it is not a high-probability event.

The final layer of vulnerability is the conflicting territorial claims in the South China sea, particularly the Spratlys, 200 tiny islands, rocks and reefs, many partially submerged with a total land area of less than three square miles. There are overlapping claims for these territories—and/or adjacent Exclusive Economic Zones (EEZs)—between China, Vietnam, the Philippines, Malaysia, Taiwan, and Brunei. Except for Brunei (whose claim is over EEZs that overlap with territories) all the parties have occupied one or more of islands they claim. These conflicting ownership claims have an impact on the 200-mile EEZs of the various claimants, and thus potentially affect energy security to the degree that the unresolved legal status of the Spratlys impedes the granting of licenses for oil and gas exploration in contested areas.

All these issues are legitimate security concerns. Since the Law of the Sea (LOS) Treaty came into force in 1980, the core ASEAN states—Indonesia, Malaysia, Thailand, Singapore, and the Philippines—have pursued significant programs of military acquisition of modern weaponry. The LOS gave them 200-mile EEZs to defend, and not surprisingly

ASEAN states began to acquire air and naval assets ranging from F-16 fighter jets to Russian MIG-29s, along with fast patrol boats, frigates, submarines, and in the case of Thailand 12,000 ton helicopter carriers.[37] The abovementioned maritime security concerns help explain why ASEAN defense budgets grew to some $16 billion before the economic crisis put acquisitions on hold. At the same time, individual ASEAN states' military cooperation with the United States steadily increased, most notably, access arrangements for American forces, including a permanent logistical base in Singapore. All the ASEAN states view the U.S. forward deployed military presence and security commitment to the region as key to regional security. But as discussed in the next chapter, there are serious reasons to doubt the proposition that security concerns, indeed, potential military conflicts, are likely to be triggered by questions of energy security. Is competition over control of resources the prime source of insecurity in contemporary Asia? Or do security dynamics reflect classic national pathologies associated with national security calculations of ambition, pride, and honor?

Chapter 9

Energy and Asian Security: Fueling Conflict or Cooperation?

A hushed tone came over the chamber as the Chinese president rose to speak to the top Party leadership. "For too long, Comrades," the President began, "China's enemies have exploited the oil riches contained in the waters around the Nansha [Spratly] and Xisa [Paracel] islands. Our scientists estimate that there are 10 billion tons of oil beneath the surface of our great southern sea. This is Chinese oil and China's 1.3 billion people need it. China is a poor, developing country and we cannot continue to import oil at the present rate demanded by the growth in our economy."

So begins a recent action thriller, *Dragon Strike,* a novel by two British journalists (and veteran Asia hands) that extrapolates from current events a scenario of global military conflict triggered by an energy-hungry China.[1] *Dragon Strike,* a genre of "faction," a mix of fiction and fact, wonderfully distills the received wisdom about the role of energy in the security calculus of an emerging, affluent, middle-class twenty-first century Asia. Asia's dynamic economic growth produces mushrooming oil demand, which, it is argued, creates energy shortages. This new wealth is also spent on military buildups. The geostrategic insecurity fueled by oil shortages, the logic continues, exacerbates territorial disputes, which erupt into military conflict over control of scarce resources.

One of the most articulate renditions of this school of thought is presented in a still-influential 1997 book, *Asia's Deadly Triangle,* by Kent Calder, a prominent Japan specialist at Princeton University subsequently appointed advisor to the U.S. ambassador in Japan. In a telling passage, he writes:

> The deadly triangle of growth, energy shortage and armament, in the context of fluid post–Cold War geostrategic alignments, threatens to destabilize

Asia, and indeed the whole Pacific. Energy shortage, the deadly but little-known link in this equation, provokes turbulence by deepening the nuclear bias of Northeast Asia, by provoking new naval rivalries centering on emerging Chinese blue-water capacity, and by deepening tensions over off-shore reserves such as those in the South China Sea.[2]

But is energy genuinely the catalyst for military conflict in Asia? As the above passage suggests there are several points at which energy and security concerns appear to intersect: territorial disputes over the Spratly Islands involving China and several ASEAN states; disputed Japanese and Chinese claims over the Senkaku/Diaoyu islands in the East China Sea; and nuclear weapons proliferation as a by-product of civilian nuclear power. Yet close inspection of all these instances where energy is viewed as a driver of Asian security dynamics suggests the perception may be flawed or greatly exaggerated by a neo-Malthusian mind-set of resource scarcity. Moreover, such an assessment fails to appreciate far more powerful undercurrents shaping the mind-set and behavior of key actors in the Asia-Pacific.

The question of energy security in Asia is, unfortunately one in which perceptions—largely based on flawed assumptions—may be having the effect of spurring both wrongheaded energy policy decisions and reinforcing trends toward confrontation in the Asia-Pacific. Fears of energy vulnerability, for example, might heighten tensions in the South China Sea, as competing claimants fear losing a prospective bonanza. China and others may accelerate the acquisition of force projection military capabilities to safeguard their respective claims to sovereignty. Or China or Japan may make unwise investments such as uneconomical pipelines from Central Asia to avoid Mid-East dependency. As discussed below, negative trends in the Asian security environment (discussed below) could easily drift towards military conflict. However, the causes have no relations to real energy security concerns. But national security decisions taken by Asian leaders may be shaped by their perceptions of prospective energy vulnerability.

Whatever else it may represent, the South China Sea controversy is emblematic of the principal issue at the heart of uncertainty about the future security of the Asia-Pacific: the meaning of China's emergence as a multidimensional great power. Will it be a revisionist power, seeking to right historical grievances by upending the status quo, or will its modernization result in a prosperous, middle-class China that acts as responsible steward of global order in which it has a large stake? Thus far, China's behavior is ambiguous. Beijing has a long track record of using force to settle questions relating to sovereignty and/or inter-state disputes since the

Communist Party took power. One careful examination of the record finds at least 118 instances between 1949–92.[3] There is a lengthy list including Korea, the Taiwan strait, Vietnam, Cambodia, the Soviet border, as well as the South China Sea.

It has been noted by one somewhat optimistic China analyst, however, that the majority of these instances of the use of force occurred prior to China's current economic reform period (most in the mid-1950s and mid-1960s) and all were related to questions of sovereignty and border or other territorial claims and none involved solely economic or resources issues.[4] One consistent thread in Beijing's behavior has been an inordinate sensitivity toward questions of sovereignty.

Against such a historical backdrop, China's behavior in the South China Sea may lend itself to an alternative interpretation. In 1992, the Chinese National People's Congress (NPC) enacted a "Law on the Territorial Waters and Their Contiguous Areas." In this legislation, China claimed sovereignty over all island territories in the South China Sea. The law restated in more comprehensive fashion long-standing Chinese claims in the area. Indeed, China had seized another South China Sea island chain, the Paracels, from Vietnam in 1974. Moreover, the bloodiest of more than a dozen minor military incidents in the Spratlys occurred in 1988 when a Sino-Vietnamese naval clash resulted in the sinking of a Vietnamese patrol boat and the death of 70 Vietnamese sailors. In the 1990s there were episodic diplomatic confrontations and military skirmishes over assertive Chinese activity on the appropriately named Mischief Reef, which lies within the Philippines' 200 mile EEZ. China first built rudimentary structures on the reef in 1995 sparking a diplomatic imbroglio with ASEAN. Then in 1998 satellite photos revealed Beijing had erected a five-story construction complete with satellite dish on Mischief Reef, which sparked an angry protest from Manila.

In terms of oil and gas, there has been a "battle of the contracts" between China and Vietnam since 1992, when China signed a contract with Crestone, a U.S. firm, to explore for oil near what Hanoi says is its continental shelf, 600 miles south of Chinese territory. Vietnam then leased exploration blocks in the overlapping claim areas to another U.S. firm. Indonesia has also pressed China on its 1992 law, under which the gas-rich Natuna Islands would fall into what Beijing says is Chinese territory. In 1995, Jakarta waited more than a year for a response from Beijing to a diplomatic query as to whether it claimed the Natunas as Chinese territory under its 1992 law. So incensed was Jakarta that it conducted unprecedented large-scale military exercises, explicitly held, "in the framework of insuring security for the development of the mega-project in the Natunas," according to an Indonesian military spokesman.[5] China finally provided an answer, telling a not entirely sanguine Indonesia that it had no claim to the Natunas' gas resources.

But how important are the Spratlys and why are they the source of such profound anxiety in the region? One oft-repeated view echoed in a report prepared for the Pentagon's Office of Net Assessment put it, "The energy potential of the Spratly islands increases the stakes for all parties."[6] The term "potentially oil rich Spratlys" is repeated mantra-like by analysts almost as one word.[7] Chinese specialists have claimed the oil reserves in the Spratly's territorial waters may be as high as 225 billion barrels, double the size of Iraq's proven reserves.[8] There have been significant oil and gas discoveries in the South China Sea, none even near the magnitude of North Sea, let alone Persian Gulf proportions. But little seismic or drilling activity have taken place in the specific area of the continental shelf where the Spratlys are located.[9] Most energy analysts consider the Spratly area gas-prone with at most modest oil potential. Veteran oil analyst Fereidun Fesharaki points out that after 15 years of offshore development and $5 billion in the East and South China Seas, peak production for China and Vietnam is estimated at 250,000 b/pd. He adds, "There is no reason to believe that a Middle East is hiding under the Spratlys."[10] In interviews with several U.S. oil company executives whose firms are active in the region, none expressed great expectations that the area would likely yield more than very modest oil reserves, and all concurred with Fesharaki's views.[11]

It is impossible to prove a negative. But there is no authoritative data on which to base the assumption of a treasure trove of oil and gas in the Spratlys that could significantly affect the energy equation of China or ASEAN claimants. Given China's likely oil import requirements to 2020 of between 6 and 9 million b/pd it is difficult to make a case that even significant finds of, say, 500,000 b/pd—which are unlikely—it would only marginally affect China's oil situation. There are other issues such as fishing rights and the strategic location of the islands astride global shipping lanes, but neither of those rises to the level of a *Dragon Strike* military scenario. Would China in effect validate the arguments of those who fear that, like Germany at the end of the nineteenth century, China is a rising expansionist power seeking to rewrite the rules of the international system on its own terms? Would it put at risk the prosperity it has built by essentially betting its future, its quest for economic modernization, on integrating itself into the global economy for such stakes?

Asian Pathologies

As we have seen in earlier chapters the whole notion of a world of resource scarcity with energy security defined in nineteenth-century neo-mercantilist terms of competition for territory and diminishing oil reserves is not the world of the first quarter of the twenty-first century. To put Asian projected oil demand growth in perspective, it might be noted that U.S. oil imports grew from 1.8 m b/pd in 1960 to 8.8 m b/pd in 1977 without creating crisis competition with other importers. Moreover, in the same time frame, oil demand for Europe and Asia rose from 4.5 m b/pd to 19 m b/pd, and by the mid-1980s world oil prices had gone into a downward spiral after the OPEC twin oil shocks of 1973–74 and 1979–80.[12] A rise in Asian oil imports from about 12 m b/pd in 1999 to 25 m b/pd by 2020 would be in a similar range, and it is not unreasonable to think it could be similarly accommodated in world oil markets. And this is a projection that does not factor in any new technologies such as the mass-marketing of hybrid or fuel cell autos likely to occur well before 2020, which would dramatically reduce oil demand. Nor do such projections consider the possibilities of new technologies that might further reduce demand for oil. More fundamentally, however, it might be asked of all claimants to the Spratlys, in the larger canvas of Asia's emergence as a major actor in the world system, why is so much importance attached to this territorial dispute? What, precisely, would be gained or lost for each country however the question is resolved?

In fact, an observer landing on earth from another planet might wonder why East Asia is so obsessed with things such as these tiny slivers of land—3 square miles total land area. It is one of the great Asian paradoxes that in an era when territory and sovereignty are being redefined by the global economy, the Internet, and spectacular technological change, long-standing territorial disputes, problems of an era past, appear to loom large as defining issues in Asian inter-state relations. Indeed, such disputes appear to be prospective triggers for conflict shaping the Asian security environment. For the past four centuries since the Treaty of Westphalia, territory has been the very definition of the nation-state. Thus, it should not be surprising if pursuing territorial claims, however puny, may be an important psychological vehicle for the collective validation of successful nation-building efforts underscored by impressive economic achievements.

Such notions may illuminate the pathos beneath another dispute over tiny uninhabited islands—the Senkaku to Japan, Diaoyu to China in the East China Sea, is no less illustrative of this paradox. The five barren rocks situated between Taiwan and Okinawa had been used as a target range by U.S. troops. When Okinawa reverted to Japanese sovereignty in 1972, the Senkakus also reverted to Japanese control, though China stakes a historical claim to them. The Senkaku/Diaoyu dispute has episodically erupted into what might be best termed political theater with a vitriolic edge. In 1996, after Japanese right-wing nationalists built a small lighthouse on the Senkakus, protests erupted in Hong Kong and Taiwan, and activists sought to sail into the islands and rip down the Japanese flag. The controversy became a magnet for vitriolic anti-Japanese sentiment in a climate of increasingly tense relations between Tokyo and Beijing. It is a measure of the depths of emotion that tempers flared no less in Hong Kong and Taiwan. This was further aggravated when Prime Minister Ryutaro Hashimoto celebrated his birthday in July 1996 by visiting the Yasakuni shrine, the burial for Japan's war dead, and then promised to return in October. The Chinese press was predictably filled with warnings of a return to Japanese militarism. "We see some people in Japan have inflated heads, and have lost their minds," opined the *People's Daily,* adding, "the people of all Asian nations must come up with a method to cool off these brains and not allow them to do anything stupid." As the frenzied rhetoric mounted, Tokyo tried to calm the outrage by withholding official recognition of the lighthouse, and canceling then Prime Minister Hashimoto's planned October 1996 visit to the Yasukuni shrine.

As with the Spratlys, rumors of oil color the dispute. Chinese oil exploration ships have occasionally ventured into territorial waters, and in the late 1990s it was not uncommon for Japanese fighter jets to scramble in response to Chinese activities in the surrounding airspace. An article in the *New York Times* during the episode said, "One complication is oil, for the East China Sea bears promising geological formations that some exports say resemble those of the North Sea."[13] While such analysis reflects the views of some energy analysts, as is the case with the Spratlys, there is no authoritative data confirming substantial oil reserves—the *New York Times* article suggested anywhere from 10 billion to 100 billion barrels of oil. Yet Western oil companies who have done substantial exploration in the East China Sea have not discovered any giant fields, have been somewhat disappointed in the area, and display little enthusiasm for further exploration.

Another of the welter of long-standing territorial disputes that occasionally have erupted into a political firestorm is that between Japan and South Korea over another small group of uninhabited islets, the Takeshima/Tokdo Islands in the Sea of Japan (East Sea to Koreans). In this instance there were not even rumors of oil. Remarkably this nasty, emotional squabble between two close U.S. treaty allies erupted in 1997 as both were facing growing threats from the proliferation of ballistic missiles and nuclear weapons from North Korea. Both the Sino-Japanese and Korean-Japanese tensions have a common underlying motif: Japan's sordid history of plunder and colonial exploitation that in varying degrees remains alive in the national memories of Chinese and Koreans.

Asia's Future Europe's Past?

While some may attribute such geopolitical turbulence in East Asia to energy security fears, there are other important factors looming beneath the surface that may better explain Asian national pathologies. In a word, it is a question of sovereignty. That is to say beneath the geopolitical uncertainty in Asia are waxing nationalisms, the collective sense of identity, narcissism, and vanity driving a new self-assertiveness. These nationalisms have begun to bump up against each other. To put them in perspective, the Spratly and Senkaku disputes discussed above are merely secondary issues in the hierarchy of potential Asian flashpoints. The two most volatile and potentially explosive disputes are essentially lingering civil wars left over from the Cold War period—North-South Korea and the cross-strait China-Taiwan problem. In both cases neither oil nor even economics enter into the calculus of conflict. A failing, anachronistic Leninist state in North Korea impedes South Korea's vision of a fully modern, globally competitive nation. Similarly for China, the Taiwan issue represents 150 years of Western and Japanese imperialism, which it seeks to overcome and achieve national unification on its course to become a great power, a status it has not achieved in the modern era.

To understand the totality of factors at play, it is worth briefly reflecting on the mind-boggling rapidity of other expressions of Asia's emergence: in the space of a single generation the region went from 4 percent of world GDP in 1960 to 24 percent by 1995. There is no shortage of telltale signs of the burgeoning technological, financial, and strategic heft of modern East Asia. Beijing recently launched its first orbiting space vehicle, prelude to a manned space program (with Russian assistance); South Korea is building a satellite launch capability. Greater China (China, Hong Kong, Taiwan), Japan, and Korea together hold some $700 billion in foreign reserves, more than half the world's total, and three of the world's ten largest economies. The 1999 earthquake in Taiwan threatened the world computer industry with semiconductor shortages, as the 1995 Kobe earthquake in Japan threatened liquid crystal computer displays.

In effect, East Asia has compressed into the time horizon of a generation a process of economic development and nation-building that occurred over nearly two hundred years in Europe. But in Europe, it took a series of wars leading first to the Congress of Vienna in 1815 before a stable system was established. Though there were conflicts such as the Crimean War and the Franco-Prussian War in the period before World War I, the international system more or less held together until it began to unravel under the pressures of competing empires in the decade before the First World War. While it is an admittedly inexact analogy, it may be instructive to recall that this the first time in modern history that China and Japan have both emerged as major economic and political/military powers at the same time. Finding a political equilibrium among two aspiring powers who are regional competitors with Confucian cultural backgrounds and developed senses of hierarchical relations has so far proved elusive. It is tempting to construct parallels between Sino-Japanese tensions and competition and that between Britain and Germany at the turn of the last century, a phenomenon wonderfully recounted in Robert K. Massie's magisterial book *Dreadnought*.[14]

There are, of course, important differences in culture, geography, and history between the European and Asian experiences and their respective civilizations. In the case of Europe it took the most horrendous mass violence in the history of this planet in two world wars before its strategic competition played out and the patterns of cooperation that have led to the European Union proved possible. In Europe, it was the assassination of Austrian archduke Ferdinand that triggered the chain of reactions that became World War I and later World War II. It is tempting to view the myriad territorial disputes and tensions in Asia as the potential contemporary analogy to Archduke Ferdinand. Beneath the triggering events in Europe was ambition and geopolitical rivalry between France and Britain—the established powers—and Germany, the rising power. In East Asia—China, and to some degree India—are contemporary rising powers. In East Asia, China and Japan may be roughly similar to France and Germany, pre–World War II as strategic rivals. Asia is unlikely to truly develop a functional sense of community until China and Japan come to terms with each other as France and Britain did with Germany in Europe—perhaps somehow balanced by the United States and a reunified Korea.

Nuclear Challenges

The lingering historic distrust, suspicion, and rivalry in East Asia assumes a new level of real—and potential—lethality with the increasingly easier access to modern weapons of mass destruction, particularly ballistic missiles. The dual nature of nuclear energy creates an unavoidable intersection of energy and security. Obtaining fissile material, which can be separated from spent nuclear fuel, is the most difficult obstacle to overcome for a potential proliferator. This is particularly sensitive in Asia, as it is the locus of civil nuclear power—as discussed above, providing one-third or more of electricity for Japan, South Korea, and Taiwan. India, China, North Korea, and potentially Indonesia also look to nuclear energy, though on a smaller scale. At present, only China is a declared nuclear weapons state under the Non-Proliferation Treaty. Yet in the real world, India is also a nuclear weapons state of still uncertain proportions.

Moreover, in the 1970s both South Korea and Taiwan were revealed to have covert nuclear weapons programs. Both were under the United States

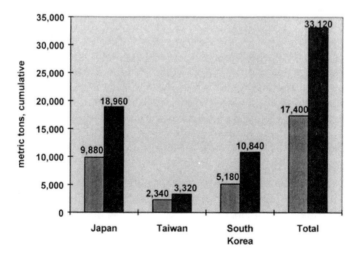

Figure 9.1 Spent Fuel from Nuclear
Source: Paul Leventhal (Nuclear Control Institute), Washington, DC, 2000.

security umbrella at the time, and the combination of U.S. security assurances and political pressure led to both abandoning their respective weapons programs. In the 1990s North Korea's efforts to acquire nuclear weapons at its nuclear complex in Yongbyon led to a crisis in 1994, one that that remains a potential threat, as since then it has developed two new generations of intermediate-range ballistic missiles.

The North Korean nuclear problem interacts with the central strategic question in the region: Sino-Japanese relations. North Korea's nuclear weapons program, combined with its missile program, has had a remarkable psychological impact on Japanese views on security. The August 1998 firing of a Taepo Dong missile over Honshu island, the heart of Japan, was a shock much like the Soviet launch of Sputnik in 1957 was to the United States. It was the first time since 1945 that any such weapon was aimed at Japan, and has fostered a more assertive attitude in Japan about defending against security threats. This in turn raises questions about how such new Japanese perspectives will impact Sino-Japanese relations.

Under the U.S. security umbrella, Japan has refrained from pursuing nuclear weapons. But it has an advanced and very sophisticated civil nuclear power program including the development of the full nuclear fuel cycle, which can provide ample fissile materials. This has led many to conclude that Japan is a "virtual" nuclear power, able to attain nuclear weapons capability rapidly if it chose to do so. At the same time, relations with China have been tense in the post–Cold War period. This is the first time in modern history that both China and Japan have emerged as multidimensional major powers at the same time. Both seek to play major roles in Asia, neither wants to be subordinate to the other, and they have yet to come to terms with each other in a broad geopolitical sense. It is not difficult to imagine Japan developing nuclear weapons should its confidence in the American security guarantee diminish. While the fissile material for nuclear weapons is the by-product of nuclear power, at bottom this is not a problem about energy. Rather, this is a problem generated by distrust and insecurity.

What developments must occur before sufficient trust and confidence is fostered in Asia, before the region attains a sense of stakeholding community, is an intriguing question. It is tempting to point to Asia's export-orientation, cutting-edge hi-tech economies, overwhelming emphasis on economic growth, and integration into the global economy as powerful forces superseding those of nationalism. Certainly, this is the logic behind the creation of Asia's fledging attempts at regional institutions, APEC for trade and the ASEAN Regional Forum for security matters. East Asian actors suspended their angst and suspicion long enough to hold an unprecedented November 1999 ASEAN+3 (Japan, China, South Korea) meeting, where they envisioned, however fanciful, a European-like future of East Asian monetary and trade unions. One test of the possibilities over the next decade or so will be the degree to which Asian states are willing to cooperate in building oil and gas pipelines, as these would link energy security in some measure.

But again, the parallel with Europe's past is worth revisiting. The turn of the last century was similarly a time of rapid technological change. The advent of transcontinental railways, steamships, and the telegraph rapidly expanded the pace and scope of commerce and finance, fostering a sense of interdependence. The United States, then accounting for one-third of the world's economy, was heavily dependent on European finance. So too were colonial enterprises in Asia, whether Malay rubber or Indian tea and textiles. The level of trade and financial integration by the turn of the century was comparable to that of the world in the last decades of the twentieth century: United States trade as a percentage of GNP did not surpass that of 1880 until 1980.[15]

These economic realities reinforced a trend in post-enlightenment thinking. As Thomas Paine put it, "If commerce were prepared to act to the universal extent it is capable, it would extirpate the system of war."[16] There was a belief, reflected in Immanuel Kant's "perpetual peace" notion, that replacing monarchies with republics would mean more accountable governments less likely to start wars. A half century later, John Stuart Mill, in his classic *Principles of Political Economy,* echoed such optimism: "Commerce, is rapidly rendering war obsolete, by strengthening and multiplying the personal interests which act in natural opposition to it. . . . The great extent and rapid increase of international trade . . . [is] the principal guarantee of the peace of the world."[17] Such views of the political potency of interdependence, particularly as global finance mushroomed by the early twentieth century, became raging conventional wisdom. This was demonstrated by the reception of Norman Angell's 1910 best-seller, *The Great Illusion,* which persuasively argued that the pervasiveness of international credit was so immense that war, which could shatter, it was unimaginable.

East Asia's remarkable prosperity creates the temptation replicate to such arguments: would societies who have worked so hard to move from village to high-rise risk it all? Moreover, it is one thing to fight over Africa, Poland, Alsace-Lorraine, particularly at a time when geography meant real power. Even Asian conflict a century ago was over Manchuria, Korea, Taiwan. But over tiny rocks in the ocean? Clearly tensions over territorial disputes involving marginal real-estate are a reflection of deeper anxieties or ambitions. Yet there are contradictory trends in both the direction of conflict and the direction of cooperation that could lead the region to follow either path.[18]

Asia's geopolitical turbulence has produced analyses, forecasts, and predictions about the future of the region. The rise of China, alternatively, a resurgent and/or collapsing Japan, and a retreating United States have all been prominent themes of assessments tending to envision, as prominent political scientist Aaron Friedberg has suggested, that "Asia's future is Europe's past." Paul Bracken has stressed the burgeoning missile, nuclear, and other modern military capacities of Asian actors to further underscore the potential both for conflict and challenge to the West, particularly new vulnerabilities of forward-deployed U.S. military bases In addition to threats from ballistic missiles and weapons of mass destruction, the disproportionate American role in maintaining stability in the Persian Gulf and East Asia has undergirded stability in both regions. The U.S. Navy's enduring commitment to enforcing the freedom of navigation has been key to the safety of the sea lines of communication (SLOCs).

Moreover, it has been the sustained U.S. role as regional balancer with 100,000 troops forward deployed overseas and a network of bilateral military base access arrangements that has fostered a relatively static security environment in Asia. But three potential major developments loom on the horizon, any one of which holds the potential to dramatically alter the geopolitics of the Asia-Pacific: the reunification of Korea; a China-Taiwan conflict; and U.S. deployments of national missile defenses. In ways that defy political science theory, Asia is bracing for change by both bandwagoning with the United States and at the same time balancing against what it sees as either unreliable or capricious American behavior, compensating for fears that the United States may gradually withdraw from the region, or perhaps as an expression of its desire for national and, collectively, regional self-assertiveness.

This has been a growing trend since the mid-1990s, evident in increasingly sophisticated hedging strategies on the part of all Asian actors.

- China has articulated a vision of the Asian future absent American alliances and in which there is a greatly diminished American role.[19] While Beijing seeks a dominant role in the region, at the same it has bet its future on integration into the global economy and has increasingly participated in global (e.g., WTO, NPT,

CTBT) and regional (e.g., ARF, APEC) institutions. Yet even as it seeks cooperative relations with the United States, its conventional and nuclear force military modernization and planning centers on attaining capabilities to prevail in a Taiwan conflict in which the United States intervenes;
- Japan's national desire to carve out its own identity as a major power in its own right (e.g., Asian monetary fund initiative, increasingly independent defense capabilities, constitutional revision debate) is accelerating. Its desire to attain a separate satellite reconnaissance ability, air refueling, and other independent defense capacities underscore this tendency. Japan also seeks to play more of a leadership role in the region, even as it simultaneously enhances the U.S.-Japan security alliance;
- South Korea's remarkable middle-power diplomacy, evidenced in the unprecedented January 2000 visit of Chinese defense minister Chi Haotian and South Korea's complex relationship with Beijing. This, even as Seoul deepens its political and military relationship with Japan in unprecedented ways and pursues satellite and ballistic missile programs even as it remains dependent on the United States for its core security.
- North Korea has managed to be at once charity case, continuing threat to, and new partner of, the United States. While developing ties to the United States, it has maintained its conventional military assets and continued to develop its weapons of mass destruction and delivery systems. At the same time, the DPRK has renewed its ties to China and Russia (witness Russian foreign minister Ivanov's February 2000 trip to Pyongyang), while expanding ties to Italy, Australia, and the Philippines and renewing normalization talks with Japan;
- The smaller and middle powers, particularly ASEAN states, have a collective desire for East Asia to be a more independent actor carrying more weight in the councils of global governance, even as they depend wholly on the U.S. military presence and commitment to assure security and U.S. markets for exports. The November 1999 ASEAN+3 meeting, replete with fantastic visions of EU-type regional arrangements for a free trade area and monetary union, was an expression of this hope, as was support for an Asian Monetary Fund. Similarly ASEAN military modernization, the acquisition of MiG-29 and F-16 fighter jets, submarines, and other naval capabilities are also illustrative of their anxieties and fears animating cooperative behavior.

A dizzying array of summitry—Sino-Russian, Sino-Japanese, Korean-Japanese, East Asia-Europe—is another manifestation of this trend. The burgeoning Asian-Middle East energy nexus discussed in chapter 4 is one example of a set of Asian interests that may be less convergent with those of the United States than in the past. Trans-Pacific trade is, of course substantial, with some $530 billion in two-way trade in 1999. The trajectory of intra-Asian trade and investment patterns are another important trend that is likely to have a crosscutting impact on U.S. interests. While the United States remains the first or second largest market for most East Asian nations, in relative terms, East Asian trade with the United States is a diminishing portion of the region's trade. In 1990, the U.S. market accounted for a third or more of Korean, Japanese, Taiwanese, and ASEAN exports. That has gradually diminished to the 20–25 percent range, with the exception of China, which is dependent on the U.S. market for more than 35 percent of its exports.

The point of this discussion of Asian security dynamics is not to argue that the world is entering a Kantian state of perpetual peace. Asia is a turbulent place. In addition to Sino-Japanese, inter-Korean, and China-Taiwan questions, there is also the prospect of nuclear competition between India and Pakistan raising the price of conflict in South Asia. Indeed, there may well be numerous horrendous military conflicts looming in Asia's future along some of the fault lines discussed above. But the source of conflict in Asia, as we have seen, is not energy shortages, though energy could be a pretext for behavior from conflict over disputed territory to the acquisition of blue water navies.

Energy does not enter into the calculus in the major flashpoints, the Korean Peninsula, the China-Taiwan question, or India-Pakistan. Even in the Spratlys, numerous discussions with a large cross section of Chinese officials and government think-tank analysts in the late 1990s suggest that sovereignty looms far larger in China's policy logic than prospective oil resources. In a moment of candor, a senior PLA officer told the author privately, "We know there is not a great pool of oil in the Spratlys, that's just the spin."[20] The marked increase in the flow of tanker traffic from the oil fields of the Persian Gulf to the ports of Pusan, Shanghai, and Tokyo does not argue for interdiction of the shipping lanes. China would have little interest in disrupting its own imports and energy flows. But raising the specter of energy security, identifying disruptions of sealane traffic provides a rationale for a new naval mission. Indeed, stirring tension involving disputed island territories can serve to reinforce support for building a blue water navy.

The key question is whether, and for how long, China will be willing to remain a "free-rider" on U.S. naval dominance as its Persian Gulf oil imports increase to substantial volumes in the 2010–2020 period. But such dependence—even if it grows from common interests—is almost certain to make China, with its pride and great power aspirations, rather uncomfortable.

As in many areas, China's behavior is ambiguous. This ambiguity is reflected in a paper on SLOC security by a prominent Shanghai-based analyst, Ji Guoxing. He writes, "China attaches much importance to SLOC security and would play a positive role in safeguarding of SLOC together with other regional countries." He adds that "the on-going naval build-up by regional countries would be another threat to SLOC security." Yet he also explains that in the mid-1980s, China changed its naval strategy from an emphasis on coastal defense to offshore defense. He cites the naval modernization plan of then chief of naval operations, Liu Huaqing, as defining "The outmost defense approaches of the Chinese navy will be spanned around the China seas: to the Korean Strait in the North, to Liuqiu Islands in the east, and to Nansha (Spratly) Islands in the south."[21] The question is whether such limited ocean-going capacity would spark a Japanese or Indian naval arms race with China. In any case, while China's military modernization promises to begin limiting American freedom of action in the decades ahead—and could even do significant damage in a Taiwan conflict in the 2004–2010 period—there is no obvious candidate to displace American naval preeminence as the guardian of the major sea lines of communication before the second quarter of this century if current trends continue.

Moreover, U.S. Navy–funded studies of the SLOCs and the key chokepoints for shipping suggest that the likelihood of disruption is low, that intentional disruptions are not likely to endure more than a matter of weeks, and that alternative routes exist that, though they raise the cost of trade, are manageable, particularly if as is highly likely, the time frame is limited.[22] A study published by the U.S. Center for Naval Analysis (CNA), *Chokepoints,* examined the question of disruption of shipping through the straits of Southeast Asia—Malacca, Sundra, Lombok (the most heavily trafficked), Phillip Channel and Strait of Singapore. It concluded that "At present, events that could disrupt passage through the Southeast Asian sealanes for an extended period of time are not likely to occur."[23] The study examined potentially human-induced closures (as opposed to say, an accidental oil spill) as a result of conflict in the South China sea, for example, or the Taiwan strait closing the sealanes, in these cases, closing the SLOC near the

Spratly or the Taiwan strait would mean diverting shipping to the Sunda and Lombock straits.

If all the straits were closed, nearly half the world's fleet, particularly oil tankers would have to divert to much longer and costlier routes, with shipping, insurance, and freight rates significantly raised. But the likelihood of all being closed in the absence of a global conflict is rather low. The CNA study concentrates on the economics of diverting shipping and suggests that a SLOC-closing scenario could be costly, but unlikely to be protracted in duration, and difficult, but economically likely to be manageable. "Economic and political interests, and geography, on balance, should work to keep open strategic straits, the study argues."[24] Given the life-and-death common interests in sustaining the flow of goods to and from East Asia, overriding common interests of all concerned parties "should work to maintain a consensus in favor of commercial freedom of navigation, the study concludes." As for the foreseeable future, the U.S. Navy is the only military power with the force projection capability to force and sustain a closure. This underscores the U.S. security guarantor role in the Asia-Pacific. It may also foreshadow the potential dangers of a naval arms race if China, Japan, and India pursue large-scale blue water navies based on the SLOC safety rationale. As nearly half the world's trade passes through the Straits of Malacca, it is difficult to envision scenarios where a major power would seek to shutdown the Malacca or surrounding straits outside of a World War II-type protracted conflict.

Energy as an Integrative Force

As evidenced in the discussion of Asian approaches to energy, notions of scarcity appear increasingly relegated to the dark recesses of the minds of policy-makers. The experience in the quarter century since the oil crises of the 1970s suggests the most probable dangers to energy security are short-term disruptions rather than questions of physical access to supply. A revolution in Iran or Saudi Arabia, for example, could disrupt flows and require displacement costs. In regard to policy responses, this suggests strategic oil stockpiles as the first line of defense, and that diversity of supply in globalized, instantaneous electronic markets may matter less, but may still provide a sense of reassurance. That is not an argument for simply letting markets take their course, though it does argue strongly for allowing markets to work. Scarcity is not the only animating force that should shape energy policies. Increasingly, scarcity is the wrong assumption on which to base policy. Markets have solved many energy policy problems that consumed governments in the 1970s. But governments shape economic environments, level the playing field, and in some instances serve as partners for and/or catalysts to private sector initiatives. One important instrument of energy security may be foreign investment, whether it is Japanese firms in Azerbaijan, Chinese firms in Sudan, or Korean oil interests in Vietnam. Efforts to decrease dependence on Middle East oil are not only measured in far-flung oil and gas equities, but increasingly in regional cooperative ventures.

The contention of this analysis is that rather than being a source of conflict, energy has the capacity to become an integrative force, creating a larger sense of shared interests and stake in cooperation. Precisely because of the strategic economic and national security implications of energy, the willingness of a nation to voluntarily link its energy fate to others in the form of a pipeline or other form of interdependence requires a certain modicum of trust and confidence. Thus, pipelines to India via Pakistan are improbable in the foreseeable future. But what of Sino-Russian or Japanese-Russian energy cooperation? In some respects, this positive national security-enhancing dimension of energy security is already beginning to be visible.

There are a number of modest but not unimportant precedents for such cooperative activities This is evidenced in the Thai-Malaysian Joint Development Area, in the Timor Gap treaty, and in Malaysian-Singapore pipelines. While these are driven by market forces, public policy in the respective capitals had to remove the obstacles to allow them to move forward. Similarly to the degree the ASEAN energy grid eventually becomes a reality, it will be largely a result of governments removing obstacles.

In Northeast Asia, there are faint outlines of similar patterns slowly taking shape. Much of it revolves around oil and gas in the Russian Far East and Siberia. Japanese investment in Sakhalin will almost certainly be the first international pipeline in the subregion. There are four Sakhalin oil and gas (both pipeline and LNG projects in various states of development) that are aimed at providing energy to Japan, China, and possibly Korea. Sakhalin II began modest oil shipments in 1999. The large amounts of gas, proximity to the Japanese markets, and Tokyo's need for clean alternatives to now postponed plans to expand nuclear energy suggest a Sakhalin-Hokkaido gas pipeline is, by the 2005–2012 period, a likely development. Western Siberia gas may hold long-term potential for China. Moreover, there are hopes of developing Irkutsk gas fields and those in Yakutia to China as well as Japan and Korea. Yet this idea has been under consideration—and the subject of numerous memoranda of understandings—for more than a decade. There are a number of possible routes that have been proposed, some traversing Mongolia, North Korea, China, and on to Korea and Japan.

Whatever the routes selected, it is difficult to envision any of the pipelines—which would be in excess of 3,000 kilometers—being feasible without significant Japanese and Korean financial support. The powerful interest by all parties and groundwork being laid for the projects suggests it is on the Northeast Asia energy agenda. However, finding the means to extract natural gas from the Russian Far East into China, Korea, and Japan is unlikely to be realized in the 2000–2010 time frame. Nonetheless, the realization of such multinational pipelines would be an important statement not only about energy security but of Sino-Russian-Japanese relations and of geopolitical stability in the region.

In the period to 2010, the most feasible Russian gas project in economic terms appears to be a Sakhalin-Hokkaido pipeline or LNG arrangement. Compared to Yakutia or Irkutsk, the nature and amount (up 65 trillion cu. ft) of Sakhalin gas reserves as well as the relatively short distance—about 620 miles (roughly 1000 km) make it the most attractive. Problems of terrain and whether complicate the prospects and lengthen the timetable for realizing either Sakhalin gas projects, as does Japan's need for multi-billion investment in its own gas infrastructure in order to be able to expand gas use. By the end of 1999 $1.3 billion total had been invested in the Sakhalin I and Sakhalin II projects. Though led by American firms, Japanese concerns have a substantial portion of both projects. One recent authoritative study projects the cost of Sakhalin gas as competitive with that from some of Japan's current suppliers, Indonesia and Australia.[25] If this project is realized in the 2010 timeframe, it could serve as an important precedent for multi-national energy cooperation and as an impetus for further Russian gas development linking Northeast Asia and the Russian Far East.

Conclusion: Policy Recommendations

There are a host of policy initiatives that logically flow from Asia's emerging energy predicament, some Asia-specific, some for U.S. policy toward Asia. Conceptually these are important because they suggest a different paradigm than current institution-building efforts in Asia. Efforts such as the ASEAN Regional Forum (ARF) and APEC are examples of what might be called ethereal multilateralism. That is to say rather than institutions such as NATO, for example, they are not built on clear and specific missions. Rather, the ARF is a vague dialogue with an emphasis on the value of just having a process. APEC has also tended to be informal, the locus of its issues economic in scope. Energy initiatives discussed below are examples of functional multilateralism, efforts to construct institutionalized forms of cooperation based on problem-solving, or focused on specific issues:

Strategic Petroleum Reserves: First, burgeoning imports put Asia in a similar position to the challenges faced by the West in the 1970s. The response then was the IEA and oil stockpiling arrangements. Experience suggests that short-term disruption is the most probable risk. But only Japan and Korea have oil stockpiles, 156 days and 90 days, respectively. It would be an important step toward greater Asian energy security to create either an Asian version of the IEA (which is part of the OECD) or a mechanism under the auspices of APEC or the ARF. In light of their experience with stockpiling, Japan and Korea, working with the IEA, might call for an initiative to bring Asian consumers up to the IEA standard of 90 days stockpiling, perhaps in phases: China and India attaining IEA levels of 90 days by 2003, Southeast Asian consumers by 2005.

Sea-lane Security: Given the perennial concern about the security of the SLOCs, and the potential drift toward naval competition, Japan and Korea might take the lead, working closely with the United States in offering collective security guarantees, particularly aimed at China and India, but to others in the region. The logic is that freedom of navigation is a universal vital security interest, and the disruption of the shipping lanes in the Straits of Malacca or elsewhere in Southeast Asia would be the economic equivalent of a weapon of mass destruction. Therefore, either as an initiative in APEC or the ARF, the respective Asian actors would commit themselves and request the United States, China, and Russia (who uniquely have the ability to enforce it) to take to the UN Security Council a resolution to guarantee access with the threat of Chapter 7 action under the UN charter to any state or non-state actor that violates freedom of access.

Maritime Piracy: In April 2000, a group of East Asian states met at the request of Japan to discuss organizing cooperative maritime action against piracy. Japan would utilize its coast guard rather than military forces to ameliorate political sensitivities. The United States would be well advised to actively support such efforts. Piracy, including hijacking of oil tankers as well as other commercial shipping are a continuing problem in the shipping lanes of Southeast Asia. This problem offers a common interest upon which to base cooperative endeavors on monitoring and responding to threats to maritime safety.[26]

Nuclear Cooperation: In the nuclear area, possibilities for fashioning an Asian version of EURATOM to manage common problems also may be an idea whose time is coming. There are a host of important issues that could be the basis of an institutional arrangement open to any Asian nation with civilian nuclear power that belongs to the IAEA. Key issues include improving nuclear safety, an early-warning system or monitoring system for radioactivity from a nuclear accident that goes beyond national borders, and managing nuclear waste. Long-term management of nuclear waste is one of the biggest challenges facing the nuclear industry. All the above-mentioned nuclear issues, most prominently, managing nuclear waste are matters of common interest to China, Japan, South Korea, and Taiwan. There is a basis for cooperation in addressing these questions, and potentially, profits for private sector firms involvement in the process. If geologically appropriate places in Russia or China for waste repositories are made available, a regional plutonium bank under the auspices of PACATOM, a proposed regional multilateral institution might begin to enter the realm of the possible.[27] PACATOM, whose agenda would be focused on the issues of civil nuclear cooperation discussed herein, along with an Asian-version of the IEA oil stockpile program, could serve as important precedents of operation multilateral security cooperation in the Asia-Pacific.

For the United States there are a host of issues that the Asian energy predicament raises. First, it should be clear that raising the profile of energy in American Asia policy greatly benefits both regional security and U.S. security interests in the Pacific. A major initiative working with the Northeast Asian nations and Russia to identify ways and means to develop Russian gas could have the added benefit of accelerating Russia's economic recovery. Similarly, development of the ASEAN energy grid could play a more central role in U.S. relations with ASEAN, the United States' fourth largest trading partner. Programs modeled on the U.S. Department of Energy's China initiative, which revolves around developing clean energy environmental technology, can be fashioned for India as well.

One area where the United States might well rethink its energy-related policies with a view toward doing less is the Caspian Basin. Rather than pressing oil consortia into premature pipeline decisions based on short-term political calculation, Washington would be well advised to let market forces shape the outcome of the various pipeline schemes. Multiple pipelines will be an important outcome, But it is far from certain that they need all be east-west. The determining factor should be the amount of commercial oil and gas available and the markets that require the products.

These are modest notional ideas to illustrate the idea that energy can be a positive, integrative factor in regional security. This analysis has tried to demonstrate that scarcity is increasingly counterproductive as a paradigm for fashioning energy security. If anything, I have probably erred on the conservative side. The array of new and emerging energy technologies discussed above is merely the tip of the iceberg. One need not be a wild techno-optimist to believe that over the coming generation forms of energy that most consumers are fully unaware of may become major energy sources—whether it is gas—or methanol-powered fuel cells, nuclear fusion, or more efficient solar energy. After all, in 1895 who would have believed the world's economy would depend on petroleum a century hence?

Notes

Preface

1. See for example, Aaron L. Friedberg, "Ripe for Rivalry: Prospects for Peace in a Multipolar World," and Richard K. Betts, "Wealth, Power and Instability: East Asia and the United States After the Cold War," in *International Security*, Winter 1993/94, vol. 18, no. 3, pps. 5–78. For the most articulate exposition of the case for a destabilizing energy-driven synergy, see Kent E. Calder, *Asia's Deadly Triangle*, Nicholas Brealey Publishing, London, 1997.

Chapter 1: Introduction

1. See for example, Julian L. Simon, *The Ultimate Resource 2*, Princeton University Press, 1996, especially epilogue, "My Critics and I," pp. 594–616.
2. See for example, Nicolas Eberstadt, "Population, Food, and Income," in Ronald Bailey, Ed., *The True State of the World*, The Free Press, New York, 1995, for discussion of the UN population "Program of Action" as an illustration of policy shaped by false assumptions.
3. See James Fallows, "Getting Along with Japan," *The Atlantic Monthly*, December 1989, pps. 34–38.
4. Robert Kaplan captured the imagination of the Clinton administration in a 1994 *Atlantic Monthly* article and subsequent book, *The Ends of the Earth*. Samuel Huntington, in a *Foreign Affairs* article later expanded into a 1995 book, *The Clash of Civilizations*, became a centerpiece of intellectual discourse.
5. Even Carson was wrong in many respects. For example, of 40 bird species she said were threatened with extinction, 33 are either stable or growing in population. For other instances where Carson was proved wrong, see Gregg Easterbrook, *A Moment on the Earth*, New York, Viking Press, 1995, especially, pp. 82–85.

6. See Al Gore, *Earth in the Balance,* New York, Houghton Mifflin, 1992, pp. 216–218.
7. For a sampling of Chicken Little warnings, see Tad Szulc, "Could This Happen to Us Again?" *Parade,* July 19, 1998. For a scholarly version, see Colin J. Campbell and Jean H. Laherriere, "The End of Cheap Oil," *Scientific American,* March 1998; and Colin Campbell, "Running Out of Gas," *The National Interest,* Spring 1998.
8. See Martin Walker, "China and the New Era of Resource Scarcity," *World Policy Journal,* Spring 1996.
9. See Joel E. Cohen, *How Many People Can the Earth Support?* New York, Norton, 1997.
10. See Paul R. Ehrlich and Anne H. Ehrlich, *The Population Explosion,* New York, Touchstone Books, 1990.
11. See H. S. D. Cole et al., ed. *Models of Doom: A Critique of the Limits of Growth,* New York, Universe Books, 1973, for a detailed analysis.
12. Ibid. p. 38.
13. See "Environmental Scares," *The Economist,* December 20, 1997, pp. 19–21, for a wonderful tour of the fallacies of eco-catastrophic forecasts.
14. Cited in Gregg Easterbrook, *A Moment on the Earth,* New York, Viking Press, 1995, 13. This is a seminal work demystifying and reevaluating environment questions and articulating a second-generation, balanced, and reasoned environmental agenda.
15. The report, prepared by the CIA Directorate of Intelligence, Office of Political Research, is dated August 1974. It was obtained initially by the *New York Times,* and also by the author, then a journalist, in 1975.
16. See Gregg Easterbrook, previously cited, especially chapter 2.
17. Quote from Nicholas Eberstadt, "Population, Food, and Income: Global Trends in the Twentieth Century," in Ronald Bailey, ed. *The True State of the Planet,* New York, The Free Press, 1995.
18. Cited in Julian Simon, *The Ultimate Resource 2,* Princeton University Press, 1996. See chapter 23.
19. Ibid. See chapters 22–23.
20. See Max Singer, "The Population Surprise," *The Atlantic,* August 1999, pp. 22–25.
21. See Peter G. Peterson, *Gray Dawn,* New York, Times Books, 1999, for a detailed assessment of the implications of the demographic shift toward older populations.
22. Cited in Nicholas Eberstadt, previously cited.
23. Nicholas Eberstadt, previously cited, p. 28.
24. Amartya Sen, *Development as Freedom,* New York: Knopf, 1999.

Notes 211

Chapter 2: The Myth of Energy Scarcity

1. See IEA, World Energy Outlook, 1998.
2. For analytical purposes, Asia refers to India as well as East Asia (China, Japan, two Koreas, and the ASEAN countries).
3. Cited in Julian Simon, *The Ultimate Resource 2*, Princeton University Press, 1995, pp. 165–66.
4. For a complete and eminently readable, panoramic history of the oil industry, see Daniel Yergin, *The Prize*, New York, Touchstone Books, 1991, also made into a PBS series.
5. Quoted in ibid., p. 626.
6. Ibid., p. 718.
7. "Oil and National Security," brief, Petroleum Industry Research Foundation, Inc., May 1999.
8. See Philip Verleger Jr., *Third Oil Shock: Real or Imaginary? Consequences and Policy Alternatives*, Institute for International Economics Policy Brief #00–04, Washington, D.C., April 2000.
9. The 1998 World Energy outlook forecasts an average of $17 a barrel to 2020, and the U.S. Energy Department's EIA also forecasts prices in that range.
10. See William Martin, "Twin Challenges: Energy and Environment in Asia," *Harvard International Review*, Summer 1997 pp. 28–31.
11. The most articulate exposition of this logic is displayed in Kent Calder, *Asia's Deadly Triangle: How Arms, Energy and Growth Threaten to Destabilize the Asia-Pacific*, London, Nicholas Brealy Publishing, 1996; in the United States, published as *Pacific Defense*, New York, William Morrow, 1996.
12. For a sampling of such near hysteria, see Tad Szulc, "Could This Happen to Us Again," Parade Magazine, July 19 1998. For a scholarly version, see Colin J. Campbell and Jean H. Laherriere, "The End of Cheap Oil," *Scientific American*, March 1998, pp. 78–83. Also James J. Mackenzie, "Heading Off the Permanent Oil Crisis," *Issues in Science and Technology*, Summer 1996.
13. See Howard Banks, "Cheap Oil: Enjoy it While it Lasts," *Forbes*, June 15, 1998, p. 84.
14. Peter Beaumont and John Hooper, *Observer* (London), July 26, 1998.
15. See Richard J. Lugar and R. James Woolsey, "The New Petroleum," *Foreign Affairs*, January-February 1999, pp. 88–102.
16. See BP 1998, Review of World Energy, June 1999, considered one of the most authoritative sources.
17. See "World Energy Outlook 1998," OECD, Paris. Also see Michael C. Lynch, "Facing the Elephant: Oil Market Evolution and Future Oil Crises," MIT Center for Energy and Development, Occasional paper #30, 1998.
18. Ibid.
19. U.S. Geological Survey, News Release, March 22, 2000.

20. Colin Campbell's projections are for oil production to peak around 2003, and then gently decline over the following 70 years. See *The Coming Oil Crisis,* Brentwood, England, Multi-Science Publishing and Petroconsultants, 1997.
21. See Campbell and Laherrere, "The End of Cheap Oil," *Scientific American,* March 1998, previously cited.
22. Ibid.
23. Cited in IEA World Energy Outlook, 1998.
24. BP Amoco Statistical Review of World Energy, June 1999.
25. C. J. Campbell, "Running Out of Gas," *The National Interest,* Spring 1998, p. 51
26. Ibid.
27. For a modern rendition, see Cutler J. Cleveland and Robert K. Kaufmann, *Energy and Resource Quality: The Ecology of Economic Process,* New York: Wiley, 1986.
28. See Michael C. Lynch, "Crying Wolf: Warnings About Oil Supply," Council on Foreign Relations Asian Energy and Security Study Group, March 1998, for a detailed critique of the methodology behind Campbell's flawed predictions, which have been wrong for more than a decade and remain so.
29. See M. A. Adelman, *The Genie Out of the Bottle,* Cambridge, Massachusetts, MIT Press, 1995, p. 2.
30. BP Amoco Statistical Review of World Energy, June 1999.
31. See Hillary Durgin, "Oil Companies Increase Exploration Spending, *Financial Times* (London), August 4, 1999.
32. See for example, Julian Simon, *The Ultimate Resource 2,* previously cited, especially chapters 11, 12, and 13.
33. See M. A. Adelman, pp. 15–17.
34. BP Amoco Statistical Review of World Energy, 1999.
35. See Sarah Emerson, "Resource Plenty: Fears of an Oil Crisis Are Misinformed, *Harvard International Review,* Summer 1997, pp. 12–15, also pp. 64–66.
36. BP Amoco Statistical Review of World Energy, June 1999.
37. See M. A. Adelman and Michael C. Lynch, "Fixed View of Resources Creates Undue Pessimism," *Oil & Gas Journal,* April 7, 1997. See also, "The Analysis and Forecasting of Petroleum Supply Bias," in Dorothea H. El Mallakh, ed., *Energy Watchers VII,* International Research Center for Energy and Economic Development, 1996.
38. Cited in Michael C. Lynch, "Crying Wolf: Warnings About Oil Supply," Council on Foreign Relations Study Group, March 1998.
39. See BP Amoco Statistical Review of World Energy, June 1999.
40. Lynch, "Crying Wolf," previously cited.
41. Ibid.
42. See World Energy Outlook, IEA, 1998 Edition, OECD, Paris. It forecasts an average price of $17 a barrel to 2010, and $25 to 2020 (in 1990 dollars).

The U.S. Energy Information Agency (EIA) projects similar prices in this timeframe.
43. Jaffe and Soligo, "China and Long-Range Asian Energy Security," Baker Institute Working Paper, April 1999.
44. See Roger N. Anderson, "Oil Production in the Twenty-first Century," *Scientific American,* March 1998, pp. 86–91, for a graphic illustration and discussion of new oil technologies.
45. Much of this analysis of the impact of technology is derived from the Rice University Baker Institute Study, "Emerging Technology in the Energy Industry," April 1999.
46. See Roger N. Anderson in *Scientific American,* previously cited.
47. John Kennedy, Editor in Chief, Oil & Gas Journal, in a speech at the James A. Baker III Institute for Public Policy, December 1998.
48. Cooper, Christopher, "Odd Reservoir Off Louisiana Prods Oil Experts to Seek A Deeper Meaning," *Wall Street Journal,* April 16, 1999; also, Roger N. Anderson, "Recovering Dynamic Gulf of Mexico Reserves and the U.S. Energy Future," *Oil & Gas Journal,* April 26, 1993, pp. 85–92.
49. See Tamsin Carlisle, "Oil Giants See 'Gusher' in Alberta Sands," *Wall Street Journal,* August 4, 1998. See also Richard George, "Mining for Oil," *Scientific American,* March 1998, pp. 84–85.
50. See Safaa A. Fouda, "Liquid Fuels from Natural Gas," *Scientific American,* March 1998, pp. 92–95; also, Curtis Rist, "Why We'll Never Run out of Oil," *Discover,* June 1999.
51. See M. Agee, "Convert Natural Gas into Clean Transportation Fuels," *Hart's Fuel Technology and Management,* vol. 7, no.2, pp. 69–72, March 1997.
52. See Tom Koppel, *Powering the Future,* Ontario, Wiley & Sons; 1999, for a discussion of Ballard Energy Systems, the pioneering company commercializing fuel cells.
53. See Richard J. Lugar and R. James Woolsey, "The New Petroleum," *Foreign Affairs,* January-February 1999, for a discussion of the possibilities for cellulosic biomass as an alternative to petroleum.
54. See Keith Bradsher, "U.S. Auto Makers Showing Interest in Fuel Efficiency," *New York Times,* January 5, 1998, p. 1.
55. Steve Licsman, "Texaco Appears to Moderate States on Global Warming," *Wall Street Journal,* May 15, 2000.
56. See Baker Institute Study, "Emerging Technologies in the Energy Industry," Rice University, Houston, April 1999.
57. Keith Bradsher, *New York Times,* previously cited.
58. See Malcolm W. Browne, "Power line Makes Use of a Miracle of Physics," *New York Times,* November 3, 1998, p. D1; see also Baker Institute energy study, previously cited.
59. Baker Institute Energy Study, previously cited.

60. See Michel May, "Energy and Security in East Asia," Institute for International Studies, Stanford University, January 1998, p. 2.

Chapter 3: The Myth of Caspian Great Game and the "New Persian Gulf"

1. See for example Jeffrey Goldberg, "Getting Crude in Baku," *New York Times Magazine,* October 4, 1998, p. 2; and Tad Szulc, "Could This Happen To Us Again? *Parade Magazine,* July 19, 1998, pp. 6–8. Szulc invokes a neo-Club of Rome argument that the world is running out of oil, and that "Much of the future of the world's economy . . . may hinge on who gains access to this wealth."
2. S. Frederick Starr, "Making Eurasia Stable," *Foreign Affairs,* vol. 75, no. 1, January-February 1996, pp. 80–92.
3. Ian Bremmer, "Oil Politics: America and the Riches of the Caspian Basin," *World Policy Journal,* vol. 15, no. 1, Spring 1998.
4. Goldberg, "Getting Crude in Baku," previously cited, p. 53.
5. For authoritative statements on U.S. policy, see Strobe Talbott, "A Farewell to Flashman," speech given at the Central Asia Institute, Washington, D.C., 21 July 1997; Stephen Sestanovich, testimony before the Senate Foreign Relations Committee, 8 July 1998; and Jan H. Kalicki, "US Strategy in the Caspian: Solution to the Caspian Puzzle," speech given at the conference "Caspian Pipelines," Washington, D.C., 19 November 1997.
6. Zbigniew Brzezinski, *The Grand Chessboard,* New York, Basic Books, 1997, particularly chapter 5.
7. Hugh Pope, "US Report Says Caspian Oil Deposits Might Be Twice as Large as Expected," *Wall Street Journal,* April 30, 1997, Section A, p. 10; and "A Caspian Gamble," *The Economist,* February 7, 1998.
8. Manik Talwani and Andrei Belopolsky, "Geology and Petroleum Potential of the Caspian Sea Region," *Baker Institute Working Paper,* April 1998. Also see "Caspian Oil: Not the Great Game Revisited," *Strategic Survey 1997/98,* Oxford: Oxford University Press for the IISS, 1998, pp. 22–29.
9. Ibid.
10. Interviews with geologists from U.S. companies such as Pennzoil, Amoco, UNOCAL, and Texaco, April 1997–October 1998. Those interviewed tended to favor estimates of below 100 billion barrels.
11. For further discussion of Georgia, see Ronald Suny, *The Making of the Georgian Nation,* Bloomington, Indiana University Press, 1988. On separatist movements and how they affect pipeline routing, see Sheila Heslin, "Key Constraints to Caspian Energy Development: Status, Significance and Outlook," Baker Institute Working Paper, Rice University, Houston, TX, April 1998.
12. U.S. oil company official, background interview with author, November 1999.

13. See Steve LeVine, "Consortium Finds Oil in Big Caspian Field, *New York Times,* May 6, 2000.
14. Author's interviews with National Iranian Oil Company (NIOC) officials, Central Asia, summer 1998. According to the interviewees, Iran would be likely to build a one million barrel per day pipeline from the Caspian to Kharg Island if it can find international partners. Tehran has already tendered to build a small pipeline from Neka on the Caspian coast to its Tabriz refinery.
15. Ronald Soligo and Amy Meyers Jaffe, "The Economics of Pipeline Routes: The Conundrum of Oil Exports from the Caspian Basin," Baker Institute Working Paper, Rice University, Houston, TX, April 1998.
16. See for example, Strobe Talbott, "A Farewell to Flashman: American Policy in the Caucasus and Central Asia," address to the Central Asia Institute, July 21, 1997; also Stuart Eizenstat, testimony before the Senate Appropriations Subcommittee on Foreign Operations, March 31, 1998.
17. Richardson quoted in Stephen Kinzer, "On Piping Out Caspian Oil, U.S. Insists, the Cheaper, Shorter Way Isn't Better," *New York Times,* November 8, 1998.
18. See Steve LeVine, previously cited, also David Stern, "Pipeline Contenders Eye Kazakh Oil Prospect," *Financial Times,* May 9, 2000. Also, author background interviews with oil industry officials, Houston, May 2000.
19. Author's interviews with British Petroleum (BP), UNOCAL, and Azerbaijan International Oil Consortium (AIOC) officials, Houston, TX, and New York, May, June, and October 1998.
20. AIOC company officials, interview with author, June 1999, October 1999.
21. Daniel Yergin, *The Prize,* New York, Touchstone, 1991, chapters 7 and 8.
22. Isabel Gorst and Nina Poussenkova, "Petroleum Ambassadors of Russia: State Versus Corporate Policy in the Caspian Region," Baker Institute Working Paper, April 1998.
23. "US-EU Statement on Caspian Energy," State Department, May 18, 1998; and "Decision on the South Pars Case," State Department, May 18, 1998.
24. Madeleine Albright, speech to the Asia Society, New York, June 17, 1998.
25. For a detailed discussion of growing U.S. economic and military involvement in Central Asia and Russian perceptions and responses to it, see Jan S. Adams, "The U.S.-Russian Face-off in the Caspian Basin," *Problems of Post-Communism,* vol. 47, no. 12, January/February 2000, pp. 49–58.
26. Hugh Pope, "Moscow Lures Back Central Asia," *Wall Street Journal,* May 22, 2000, p. A29.
27. Afshin Valinejad, "Group Raises Price on Rushdie's Head," *USA Today,* October 13, 1998, p. 18A.
28. See R. K. Ramazani, "Iran After the Khatemi Election," *Middle East Journal,* vol. 52, no. 2, Spring 1998, pp. 179–86; and Shaul Bakhash "Iran's Unlikely President," *New York Review of Books,* November 5, 1998, pp. 47–51.

29. Speech given by Kazak President Nursultan Nazabayev, James A. Baker III Institute for Public Policy, November 1991.
30. Martha Brill Olcott, "The Caspian's False Promise," *Foreign Policy,* Summer 1998.
31. Economist Intelligence Unit, Azerbaijan Country Profile, 1999–2000.
32. Economist Intelligence Unit, Kazakhstan Country Profile, 1999–2000.
33. Joe Barnes, "US Interests in the Caspian Basin: Getting Beyond the Hype," Baker Institute Working Paper, April 1998.
34. See Henry A. Kissinger, "The End of NATO as We Know It?" *Washington Post,* August 15, 1999, p. C7.
35. Author's interview with oil industry officials, Houston, TX, September 1998. Also see Christopher Cooper and Hugh Pope, "Dry Wells Belie Hope for Big Caspian Reserves," *Wall Street Journal,* 12 October 1998, p. A13.
36. Kortunov, "Russia and Central Asia."
37. Richard Sokolsky and Tanya Charlick-Paley, "Look Before NATO leaps into the Caspian," *Orbis,* vol. 43, no. 2, Spring 1999, pp. 285–297.
38. Yang Shuheng of the Peace and Development Institute, Beijing 1998.

Chapter 4: The Asian Energy Predicament

1. See Kent E. Calder, "Asia's Empty Tank," *Foreign Affairs,* March/April 1996, p. 56.
2. See Daniel Yergin, "Asian Energy Needs and Security Implications," paper for the International Institute for Strategic Studies, September 1997; Las Yergin, et al., "Fueling Asia's Recovery," *Foreign Affairs,* March/April 1998, pp. 34–50.
3. For purposes of analysis, East Asia refers to Japan, China, the two Koreas, the nations of ASEAN, and Australia. References to Asia or Asia-Pacific adds India to the frame of reference. Minus India, these encompass the three categories for the region used by the IEA—East Asia, China, and Pacific OECD.
4. See *The East Asian Miracle,* The World Bank and Oxford University Press, 1993.
5. See BP Amoco Statistical Review of World Energy, 1999, p. 10.
6. See Keun-Wook Paik, "Gas and Oil in Northeast Asia," London, Royal Institute of International Affairs, 1995, pp. 3–5.
7. BP Amoco Statistical Review of World Energy, 1999.
8. See Kent E. Calder, *Asia's Deadly Triangle: How Arms, Energy, and Growth Threaten to Destabilize Asia-Pacific,* London, Nicolas Brealey Publishing, 1996 (in United States, published as *Pacific Defense,* New York, William Morrow). This is most cogent articulation—and most frequently cited rendition—of the resource scarcity-territorial dispute-conflict energy security logic. See also Calder, "Asia's Empty Tank," *Foreign Affairs,* March/April 1996, or, for

example, Mamdouh G. Salameh," China Oil and the Risks of Regional Conflict," *Survival,* Winter 1995–96.
9. See IEA World Energy Outlook, 1999, OECD, Paris, June 1999 update.
10. IMF World Economic Outlook, September 1999; also *Far Eastern Economic Review,* October 7, 1999, pp. 98–99, for other estimates.
11. For an analysis of the financial crisis see Robert A. Manning, "The Asian Financial Crisis: Security Risks and Opportunities," 1998 Pacific Symposium, National Defense University, Institute of National Strategic Studies, Washington, D.C.
12. Institute for International Finance, see also G. Pierre Goad, "Opening the Money Tap," *Far Eastern Economic Review,* October 7, 1999, pp. 84–85.
13. See Michael Schuman and Jane L. Lee, "Dismantling of Daewoo Shows How Radically Korea Is Changing," *Wall Street Journal,* August 17, 1999, p. 1.
14. President Kim Dae Jung, interview with author, June 1999.
15. See, for example, David E. Sanger and Mark Landler, "Asian Rebound Derails Reform as Many Suffer," *The New York Times,* July 12, 1999, p. 1, and Peter Montagnon, "False Dawn in Asia," *Financial Times,* April 19, 1999, p. 17. See also, "Asia: How Real is the Recovery," *Business Week,* May 3, 1999; and Paul Krugman, "Has Asia Recovered?" *Time,* July 14, 1999.
16. *Petroleum Intelligence Weekly,* October 24, 1999, p. 14, and *Oil Market Report,* International Energy Agency, April 11, 2000.
17. See World Energy Outlook 1998, IEA, Paris, 1999, pp. 276–278, for a discussion of the contentious debate among economists. There is great doubt about the accuracy of official Chinese growth statistics, see also Angus Maddison, "Measuring Chinese Economic Growth and Levels of Performance," OECD, Paris, 1997. It appears that official growth figures have been overstated by 2–3 percent since 1954.
18. See "World Energy Outlook and the Impact of Economic Turmoil in Asia on Oil Prospects," June 1999, International Energy Agency. Also, IMF, "World Economic Outlook," Washington, D.C., 1999.
19. US DOE, Energy Information Agency, "International Energy Outlook," 1999, p. 141.
20. See *Northeast Asian Energy in a Global Context,* the Royal Institute of International Affairs, 1996, p. 11.
21. My calculations from IEA World Energy Outlook and EIA International Energy Outlook, 1999.
22. My calculations derived from BP Statistical Review of World Energy, 1999, and Ichizo Aoyama and Richard Berard, "The Asian Oil Imbalance 1996–2010," research paper for Persian Gulf energy study of the Baker Institute for Public Policy, Rice University, 1997.
23. My own extrapolations based on calculations in Paul Hornswell, *Oil in Asia,* London, Oxford University Press, 1997, p. 22.

24. *Northeast Asia Energy in a Global Context,* previously cited, p. 3.
25. World Automobile Association.
26. *China Energy Databook,* Jonathan Simon, Editor, Lawrence National Laboratory, 1996, and China, State Statistical Bureau, and own calculations.
27. My own calculations based on IEA and Asian Development Bank energy statistics.
28. My own calculations from BP Amoco Statistical Review of World Energy.
29. My own calculations from BP Amoco Statistical Review of World Energy; APEC Energy Advisory Committee, June 1995, cited in Kent Calder, "Energy and Security in Northeast Asia: Fueling Security," IGCC Policy Paper, University of California, February 1998.
30. My own calculations based on IEA forecast, EIA, and Fereidun Fesharaki of East-West Center (1998).
31. Chinese nuclear officials, interview with author, Beijing and Guangzhou, March 1999.
32. My own calculations based on Fesharaki, presentation to Council on Foreign Relations Energy Security Study; also background interview with U.S. Department of Energy official, June, 1999.
33. This assertion is based on numerous interviews with Chinese energy officials and Western oil and gas multinationals.
34. See the APEC website, APEC Energy Working Group for notional ideas of an ASEAN energy grid; see Paik, (1995) previously cited for a plethora of ideas for a Northeast Asia gas grid. Both ideas are still at an embryonic stage of development.
35. See EIA International Energy Outlook, 1999.
36 BP Amoco Statistical Review of World Energy, 1998.
37. BP Amoco Statistical Review of World Energy, 1998.
38. It is possible that Tarim will prove to contain 20 billion barrels of commercially extractable oil, and that the South China Sea holds 100 trillion cu. ft. of natural gas. But so far no major oil companies believe either to be the case.
39. See Paul Hornswell, "Oil in Asia," previously cited, chapter 10, pp. 251–277.
40. Ibid. Also author background interviews with oil executives from three major firms active in Asia, September 1999.
41. See Hornswell, previously cited, chapter 11, "Middle East Exporters and Asia," for a detailed discussion of oil supply relationships, particularly pricing in the Dubai forward market and the Saudi "Asian premium."
42. See Fadhil Chalabi in "Gulf-Asia Energy Interdependence," Middle East Institute, 1998.
43. EIA, International Energy Outlook 1999, and Fesharabi, et al., *Pacific Energy Outlook,* Honolulu, East-West Center, 1995.
44. Kent E. Calder, "Asia's Empty Gas Tank," *Foreign Affairs,* March/April 1996.

45. Yamani, speech to the Centre for Global Energy Studies, 8th annual conference, April 27, 1998, London.
46. Lukman speech in John Calabrese, ed., *Gulf-Asia Energy Security,* Middle East Institute, Washington, 1998.
47. See *South China Morning Post* (Internet edition), November 4, 1999, also *China Daily,* November 4, 1999.
48. See *Middle East Digest,* June 9, 1995.
49. Cited in Aoyama and Berard, "The Asian Oil Imbalance 1996–2010," working paper for Middle East and the Gulf study, James A. Baker III Institute for Public Policy, Rice University, 1997.
50. See Ronald Soligo and Amy Jaffe, "China's Growing Energy Dependence, "working paper for "China and Long-Range Energy Security," study of the James A. Baker III Institute, Rice University, April 1999.
51. *Financial Times,* November 6, 1999, p. 1.
52. *Washington Post,* July 15, 1997.
53. See Jonathan Rynhold, "China's Cautious New Pragmatism in the Middle East," *Survival,* vol. 38, no. 3, Autumn 1996, pp. 102–16, for a discussion of how China balances strategic considerations with the demands of modernization.
54. Author background interviews with Chinese officials, July 1997, and U.S. officials September 1997, February 1998.
55. See Bates Gill, "Chinese Arms Exports to Iran," *Middle East Review of International Affairs (MERIA),* Internet edition, Tel Aviv, vol. 2, no. 2, May 1998. See also Gill, "Two Steps Forward, One Step Back: The Dynamics of Chinese Non-Proliferation and Arms Control Policy-Making in an Era of Reform," in David M. Lampton, ed., *Chinese Foreign and Security Policy Decision-making in an Era of Reform, 1985–2000,* forthcoming.
56. See Geoffry Kemp, "Energy Superbowl," the Nixon Center, Washington, D.C., 1997, for a thoughtful discussion of the Gulf, Caspian, and energy security.

Chapter 5: China: In Search of an Energy Policy

1. See Kenneth B. Medlock III and Ronald Soligo, "The Composition and Growth of Energy Demand in China, "Working Paper for China and Long-Range Asia Energy Security" study, James A. Baker III Institute for Public Policy, Rice University, April 1999. They project that at an average of 5 percent annual growth, China's motor vehicle stocks could increase to 3 per hundred by 2015.
2. See "Resource Warriors, " *Asian Wall Street Journal,* July 23, 1997.
3. See David B. Ottaway and Dan Morgan, "China Pursues Ambitions Role in Oil Market," *Washington Post,* December 26, 1997, p. A35, and "Beijing

Gusher," *Far Eastern Economic Review,* February 26, 1998, for major press interpretations of China's oil strategy.
4. See Gaye Christofferson, "China's Intentions for Russian and Central Asian Gas," National Bureau of Asian Research, March 1998, for a detailed argument that the CNPC acquisitions were part of a larger strategy to minimize growing dependence on Middle East oil and gas, diversify supplies, and forge a multinational energy grid in Northeast Asia.
5. For a still relevant and authoritative discussion of energy policymaking, albeit in the 1980s, see Kenneth Lieberthal and Michel Oksenberg, *Policy-Making in China, Leaders Structures, Processes,* Princeton University Press, 1988.
6. See *Oil & Gas Journal,* August 30, 1999, for PRC decision against Kazakhstan pipeline, and Baker Institute report, "China and Long-Range Asian Energy Security," for incompatibility of Iranian crude with Chinese refineries.
7. Author interviews with Western oil officials and Chinese officials, Beijing, Chengdu, and Gaungzhou, April 1999.
8. See the IEA, *World Energy Outlook,* pp. 273–98; as noted in the previous chapter, many economic analysts believe China's growth statistics may overstated by 2–3 percent.
9. See Edward S. Steinfeld, "Will China be the Next Giant to Fall?" Nikko Capital Trends, Nikko Research Center, Washington, D.C., August 1998. See also Steinfeld, *Forging Reform in China: The Fate of State-Owned Industry,* Cambridge University Press, 1998 for a lucid critique of China's SOEs.
10. Steinfeld, Nikko Capital Trends, previously cited.
11. Cited in Nicholas Lardy, "China and the Asian Contagion," *Foreign Affairs,* July/August 1998, p. 80.
12. See Hugo Restall, "Is China Headed for a Crash," *Wall Street Journal,* September 2, 1999.
13. See James Kynge, "Report Reviews China Banking Abuses," *Financial Times,* December 17, 1999.
14. See Nicholas Lardy, "China's Unfinished Economic Revolution," Brookings Institution, Washington, D.C., for detailed analysis of China's precarious financial situation.
15. See Jiang Zemin's Report at the 15th National Congress of the Communist Party, September 12, 1997, Chinese U.S. embassy website.
16. Hugo Restall, *Wall Street Journal,* previously cited.
17. *London Financial Times,* August 14, 1999, p. 1. On corruption probe, see *South China Morning Post* (Internet edition), October 13, 1999.
18. Cited in the U.S. Energy Information Administration, "China: Environmental Issues," Washington, D.C., October 1999.
19. See Hugo Restall, *Wall Street Journal,* previously cited.
20. See Steven W. Lewis, "Privatizing China's Oil Companies," Working Paper of Baker Institute, Rice University, study on China and Long-Range Energy Security, April 1999.

21. Ibid., p. 12.
22. U.S. Energy Information Agency, Department of Energy, October 1999; also, see Ernst & Young, Petroleum Economist, "Energy Map of China," August 1998.
23. BP Statistical Review of World Energy 1999. Again, note that proven reserves only refers to resources that may be extracted with reasonable certainty from known deposits under existing economic and technological operating conditions. Changes in price or technology may yield substantially larger proven reserves in the future. This is also true for oil and gas.
24. Cited in Todd M. Johnson, "Foreign Involvement in China's Energy Sector," in Elizabeth Economy and Michel Oksenberg, eds. *China Joins the World,* New York, Council on Foreign Relations Press, 1999, p. 271.
25. See Li Peng, "China's Policy on Energy Resources," *Qiushi* [Seeking Truth] no. 11, 1997, FBIS translation, July 15, 1997.
26. See IEA, World Energy Outlook, 1998; also see Vaclav Smil, "China's Energy and Resource Uses," *The China Quarterly,* Fall 1998.
27. See China Energy Databook, Lawrence Livermore laboratory, September 1996, chapter 2.
28. Cited in U.S. Energy Information Administration, Department of Energy, "China: Environmental Issues," October 1999.
29. ARCO officials, interview with author, fall 1999.
30. U.S. Energy Information Administration, China energy profile, June 1999. See also, Vaclav Smil, *China Quarterly,* previously cited, and Smil, *Energy in China's Modernization,* New York, M. E. Sharpe, 1988. Some information confirmed to author in background interviews with Chinese officials, April 1999.
31. See "China 2020: the Development Challenges in the New Century," The World Bank, Washington, D.C., 1997.
32. See "China's Electrical Power Options," joint study by Battelle Memorial Institute, Beijing Energy Efficiency Center, and the Energy Research Institute of China, Pacific Northwest Laboratory, April 1998, Washington, D.C.
33. Department of Energy, International Energy Outlook 1999.
34. Department of Energy, International Energy Outlook 1999. Also, DOE officials in interview with author, November 1999.
35. See Department of Energy, International Energy Outlook, 1999. For a Chinese critical perspective, see Dai Qing, *The River Dragon Has Come!* New York: M. E. Sharpe, 1998.
36. Interviews with nuclear and provincial officials, Guangzhou, March 1999.
37. Interviews with Chinese officials in Beijing and Guangdong, March 1999.
38. Wang Xianzheng, a senior coal industry official said China seeks to reduce coal to 60 percent of its energy mix by 2010 and 50 percent by 2020. *China Daily,* December 5, 1999.

39. See IEA "World Energy Outlook," OCED Paris, 1999, pp. 289–90.
40. See *China Daily,* September 18,1999, p. 1
41. Guangdong provincial nuclear officials, interview with author, March 1999.
42. See Jennifer Ehrlich, "Anger Over Plan for Third Nuclear Plant," *South China Morning Post* (Internet edition), January 19, 2000.
43. See IEA, World Energy Outlook, 1998 edition, OECD Paris, 1999. See pp. 273–95.
44. Author's calculation, from BP Amoco Statistical Review of World Energy, 1998.
45. See Ronald Solito and Amy Jaffe, "China's Growing Energy Dependence: The Costs and Policy Implications of Supply Alternatives," Working Paper of the Baker Institute for Public Policy, Rice University, April 1999.
46. Author background interviews with Western oil company officials in Beijing, April 1999, and Washington, D.C., June and October 1999.
47. U.S. Department of Energy China Country profile, 1995; also author interview with Chinese officials, March 1999.
48. Cited in Hornswell, *Oil in Asia,* London, Oxford University Press, 1997.
49. "World Energy Outlook," IEA, Paris, 1999.
50. "China," country profile, Department of Energy, U.S. Energy Information Administration, June 1999.
51. Western oil company officials, interview with author, Beijing, April 1999.
52. BP Amoco Statistical Review of World Energy, 1999; Baker Institute China and Long Range Asian Energy Security Study, previously cited, and Fersharaki, East-West Center. China is putting on a quota for oil imports in 1998 and early 1999, essentially banning exports, led to confused figures (and a fair amount of smuggling), rendering precise statistics problematic.
53. Growth and import demand figures, author extrapolations from Soligo and Jaffe, "China's Growing Energy Dependence: The Costs and Policy Implications of Supply Alternatives," Baker Institute 1999, previously cited, and Fereidun Fesharaki and Kang Wu, "China's Role in the Asian Oil Market," paper presented at CGES Annual Conference, London, April 1998.
54. See *People's Daily,* Beijing, September 21, 1994.
55. See *Oxford Analytica Asia Pacific Daily Brief,* January 5, 2000.
56. Northeast Asian Energy in the Global Context, Royal Institute of International Affairs, London, 1996. See also Yang Guang, in Gulf-Asia Energy Security, Middle East Institute, Washington, D.C., 1998.
57. This figure is based on background interviews with Chinese officials and Western economists' informal estimates.
58. See John Pomfret, "Chinese Ties Leaders Smuggling," *Washington Post,* January 22, 2000. In just one smuggling scandal in Fujian Province, it was revealed 42 million barrels of oil were being smuggled into China.
59. See Trish Saywell and Ahmed Rashid, "Innocent Abroad," *Far Eastern Economic Review,* February 26, 1998, p. 50.

Notes 223

60. See Soligo and Jaffe, "China's Growing Energy Dependence," Baker Institute, previously cited.
61. See Xioaje Xu, "The Dragon's Rise," Baker Institute Working Paper, previously cited.
62. Chinese officials and Western oil company officials, interview with author, Beijing, March and April 1999.
63. See Soligo and Jaffe, "China's Growing Energy Dependence," Baker Institute, 1999, previously cited.
64. See David Blumenthal and Gray Sasser, "Fuel for the Next Century," *China Business Review*, July-August 1998, pp. 33–38; also Soligo and Jaffe, "China's Growing Energy Dependence," Baker Institute, previously cited.
65. See Baker Institute study, "China and Long-Range Energy Security, previously cited. See also "China's Oil Price Reforms A Major Step ion Deregulating Its Petroleum Sector," *Oil and Gas Journal*, August 10, 1998, pp. 46–48.
66. See Ho Swee Lin and James Kynge, "Investors to Treat Warily in China Oil Fields," *Financial Times,* December 23, 1999. See also "PetroChina Hopes to Shake Off Its Past," *Wall Street Journal,* January 13, 2000.
67. Cited in Elizabeth Economy, "Progress Inhibited in Polluted Atmosphere," *South China Morning Post,* November 22, 1999.
68. See BP Amoco Statistical Review of World Energy.
69. See Paik and Lan, "China Preps to Expand Gas Output and Distribution Amid Challenges, *Oil & Gas Journal,* July 20, 1998, pp. 27–31; see also Jeff Logan and William Chandler, "Incentives Needed for Foreign Participation in China's Natural Gas Sector, *Oil & Gas Journal,* August 10, 1998.
70. Xiaoje Xu, "The Gas Dragon's Rise: Chinese Natural Gas Strategy and Import Patterns," Baker Institute, Working Paper, April 1999. Also CNPC officials, interview with author, April 1999.
71. See Xioaje Xu paper, previously cited.
72. Senior Western oil company official, interview with author, December 1999.
73. See, for example, Keun Wook-Paik, "Gas and Oil in Northeast Asia," Royal Institute for International Affairs, London, 1995; *Oil & Gas Journal,* July 6, 1998, August 10, 1998; Gaye Christofferson, "China's Intentions for Oil and Gas in Central Asia," previously cited.
74. See Xiaoje Xu, previously cited and Keun-Wook Paik and Quan Lan, "China Preps to Expand Gas Output and Distribution Amid Challenges," *Oil & Gas Journal,* July 20, 1998.
75. See Jeff Logan and William Chandler, "Incentives Needed for Foreign Participation in China's Natural Gas Sector," *Oil & Gas Journal,* August 10, 1998.
76. See "China's Electric Power Options," Battelle Memorial Institute and Beijing Energy Efficiency Center, June 1998.
77. Author interviews with foreign and defense officials and think-tank analysts, March, April, and June 1999.

78. Barry Rubin, "China's Middle East Strategy," *Middle East Review of International Affairs,* vol. 3, no. 1, March 1998.
79. Xiaoje Xu, "The Dragon's Rise," Baker Institute Working Paper, previously cited.
80. See Samuel P. Huntington, *The Clash of Civilizations and the Remaking of World Order,* New York, Simon & Schuster, 1996.
81. See article on the report form the China Strategy and Management Society, a Beijing think tank, in Ta Kung Pao, March 15, 1999, Hong Kong.
82. Ta Kung Pao, previously cited.
83. See Soligo and Jaffe, "China's Growing Energy Dependence," Working Paper for the Baker Institute Study on China and Long-Range Asian Energy Security, April 1999, previously cited.
84. See IEA, World Energy Outlook 1998, chapter 15.
85. See Vaclav Smil, "China's Energy and Resource Uses," *China Quarterly,* Spring 1998. See also Smil, *China's Environmental Crisis,* New York, M. E. Sharpe, 1988, for a detailed analysis of China's environmental predicament.

Chapter 6: India: The Lumbering Elephant

1. GNP figures are not based on Purchasing Power Parity.
2. IEA, World Energy Outlook, 1999; also IMF statistics.
3. For a skeptical view of India's privatization prom see Dan Lefkowitz, "India's Privatization Process is Half-Hearted," *Asian Wall Street Journal,* weekly edition, September 23, 1999, p. 14.
4. See *Wall Street Journal,* October 3, 1999
5. See Sadanand Dhume," India's Awakening," *Far Eastern Economic Review,* January 20, 2000, pp. 36–40.
6. See International Energy Outlook, 1999, U.S. Dept. of Energy, Energy Information Agency; also see IEA, World Energy Outlook, OECD, Paris, 1999.
7. "New Power Policy," *India Business Intelligence,* January 1999, Economist Intelligence Unit, London. Also, "Energy in India: An Overview," *CSIS South Asia Monitor,* October 1, 1999 (Washington, D.C.), cites 18 percent peak shortages.
8. "India: Economics, Demographics, and Environment," U.S. Dept. of Energy, Energy Information Agency, 1998.
9. "The New Prize," *The Economist,* June 18, 1994, p. 4.
10. See "India: Country Survey," U.S. Energy Information Agency (EIA) June 1999. See also, V. Vasuki Rao, "India: Many Vows, But Few Volts," *Journal of Commerce,* August 22, 1997.
11. See "The Electrification of Everywhere Else," *The Economist,* June 18, 1994, p. 7.
12. IMF World Economic Outlook, 1999, author's calculations.

13. IEA, World Energy Outlook 1998; also author background interviews with Western oil industry officials, October 1998.
14. Parliamentary report cited in Gained Singh (currently Foreign Minister) "What Constitutes National Security in a Changing World Order? India's Strategic Thought," p. 10, Occasional Paper No. 6, Center for the Advanced Study of India (CASI), University of Pennsylvania, June 1998.
15. Rahul Roy-Chaudhury, "An Energy Security Policy for India: The Case of Oil and Gas," *Strategic Analysis,* February 1998, Institute for Defense Studies and Analysis, New Delhi.
16. "Energy in India: An Overview," *South Asia Monitor,* no. 14, October 1, 1999, Center for Strategic and International Studies, Washington, D.C.
17. See *India: Energy Structure,* U.S. Dept. of Energy, Energy Information Service, 1997.
18. See India, country survey, previously cited, U.S. Energy Information Agency, June 1999; IEA World Energy Outlook, 1998.
19. India: Country Survey, EIA, June 1999, previously cited.
20. See *India: Environmental Issues,* U.S. Dept. of Energy, Energy Information Agency, November 1999. Some sources, such as Ragu G. C. Thomas, put the figure for coal at closer to 60 percent, while the EIA suggests it is closer to 75 percent.
21. India: Country Survey, previously cited. U.S. Energy Information Agency, June 1999.
22. Ibid.
23. *India: Environmental Issues,* EIA, November 1999, previously cited.
24. Address by the president of India to Parliament, New Delhi, October 25, 1999, Indian U.S. Embassy website.
25. "India," Department of Energy, EIA, June 1999, previously cited. Also see "Let's Try Again," *India Business Intelligence,* January 1999, Economic Intelligence Unit, London.
26. See Patrick Crow, "Watching Government India's Changes," *Oil & Gas Journal,* April 10, 2000, internet edition.
27. For a discussion of power infrastructure financing, see R. K. Jain, "Financing of Power Projects in India: Problems and Prospects—A Note," *Reserve Bank of India Occasional Papers,* vol. 19, no. 3, September 1998. See also *India Business Intelligence,* January 1999, previously cited.
28. U.S. Dept. of Energy, EIA, International Energy Outlook, 1999.
29. See "Indian Oil Industry Doubling Efforts to Attract International Interest," interview with ONGCC official K. K. Sharma, *Oil & Gas Journal,* November 1, 1999, internet edition. See also "India's Gas Infrastructure Plan Build Steam," *Oil & Gas Journal,* July 5, 1999.
30. See "India's Power Projects Drive Boom in LNG Export Schemes," *Oil & Gas Journal,* October 4, 1999, also see Jeremy Beckman, "Long-Term Contracts Needed for Imports," *Oil & Gas Journal,* April 5, 1999, internet edition.

31. "India: Country Survey," U.S. EIA June 1999, previously cited.
32. See "India's Gas Infrastructure Plan Builds Steam," *Oil & Gas Journal,* July 5, 1999, internet edition.
33. See "India's Power Projects Drive Boom in LNG Import Schemes," *Oil & Gas Journal,* October 4, 1999.
34. Cited in Paul Hornswell, "Oil In Asia," p. 90, London, Oxford University Press, 1997.
35. See Paul Hornsnell, *Oil in Asia,* Oxford University Press, 1997, pp. 90–91, for a historical overview of India's oil industry.
36. BP Amoco Statistical Review of World Energy, June 1999, London.
37. Ibid.
38. Cited in "India: Country Survey," U.S. Dept. of Energy, Energy Information Agency, Washington, June 1999.
39. See, for example, Rahul Roy-Chaudhury, "An Energy Security Policy for India: The Case of Oil and Natural Gas," Strategic Analysis, February 1998, Institute for Defense Studies and Analysis, New Delhi. Also, oil executives interviewed by author on background, June 1998, April 1999, October 1999.
40. "Indian Oil Industry Doubling Efforts to Attract International Interest," *Oil & Gas Journal,* November 1, 1999, internet edition.
41. Cited in Shebonti Ray Dadwal, "India's Energy Situation: Crisis in the Making," *Strategic Analysis,* IDSA, New Delhi, June 1997.
42. Quoted in "India's Power Projects Drive Boom in LNG Import Schemes," *Oil & Gas Journal,* October 4, 1999, Internet edition.
43. Ibid.
44. See Shebonti Ray Dadwal, "India's Energy Situation," previously cited.
45. U.S. Energy Information Agency (EIA) "South Asia Regional Overview," April 1999. See also *Financial Times Bangladesh Survey,* May 5, 1998.
46. EIA, South Asia Regional Overview, previously cited.
47. Senior analyst, at a leading U.S. consulting firm, background interview with author, November 1999.
48. See Sujit Dutta, "Indo-Gulf Relations: Dimensions of Security," in *Gulf-Asia Energy Security,* John Calabrese, ed. Middle East Institute, Washington, D.C., 1998.
49. Rahul Roy-Chaudhury, "An Energy Security Policy for India: The Case of Oil and Natural Gas," *Strategic Analysis,* February 1998, IDSA, New Delhi.
50. Jaswant Singh, "Against Nuclear Apartheid," *Foreign Affairs,* September-October 1998, pp. 41–52.
51. Sujit Dutta in *Gulf-Asian Energy Security,* previously cited.
52. "India's Forgotten Force," *Jane's Defense Weekly,* March 11, 1995.
53. "Energy Strategies and Military Strategies in Asia," prepared for the Advisor to the Secretary of Defense for Net Assessment, Washington, D.C., September 1999.

54. See George Perkovich, *India's Nuclear Bomb,* University of California Press, 1999. See especially pp. 1–60, and 448–68.
55. See George Tanham, *Indian Strategic Thought: An Interpretative Essay,* RAND Corp. Santa Monica, 1993, for a clear analysis of Indian geostrategic perspective and worldview.
56. U.S. official with long nuclear experience, interview with author, December 1999.
57. Address by K. C. Pant, Deputy Chairman, Government Planning Commission, to the International Conference, "India in the New Millennium," Harvard University, November 8, 1999.
58. For a detailed exposé of India's nuclear safety problems, see Nayan Chanda, "The Perils of Power," *Far Eastern Economic Review,* February 4, 1999, pp. 10–17.
59. See IEA, World Energy Outlook 1998 edition.

Chapter 7: Japan/Korea: Whither Asia's Mature Consumers?

1. Kent Calder, "Asia's Empty Tank," *Foreign Affairs,* March/April, 1996.
2. BP Amoco Statistical Review of World Energy, 1999, author calculations.
3. "Energy in Japan," MITI, Tokyo, 1999. Also, Japan: Country Overview, June 1999, U.S. Energy Information Agency (EIA).
4. "Energy in Japan," Ministry of Trade and Industry, Tokyo, 1999; also Institute of Energy Economics, Japan 1997.
5. Senior Korean official, background interview with author, December 1999.
6. See Richard Samuels, *The Business of the Japanese State: Energy Markets in Comparative and Historical Perspective;* Ithaca, NY, Cornell University Press, 1987, for a detailed analysis of Japan's postwar energy policies.
7. See Paul Hornsnell, *Oil in Asia,* London, Oxford University Press, 1997, chapter 4.
8. See Clay Chandler, "Japanese Economy Shrinks Again," *Washington Post,* February 7, 2000.
9. See "Report of the Prime Minister's Commission on Japan's Goals in the 21st Century," Office of the Prime Minister, Tokyo, January 2000.
10. See IA World Energy Outlook, 1998 edition, and "Energy in Japan," MITI, Tokyo, 1999.
11. See Daniel Yergin, *The Prize,* New York, Touchstone Books, 1991, chapter 29, "The Oil Weapon."
12. "Energy in Japan," MITI, Tokyo, 1999.
13. Ibid.
14. Cited in "Japan: Country Overview," U.S. Energy Information Agency (EIA) May 1999; also, MITI, "Energy Balance Tables," 1998.

15. See Keun-Wook Paik, *Oil and Gas in Northeast Asia*, Royal Institute of International Affairs, London, 1995.
16. Ibid.
17. See Alexandra Harney, "Japan's Saudi Oil Drilling Plan Falters," *Financial Times,* January 16, 2000.
18. "Energy in Japan," MITI, Tokyo, 1999.
19. "Energy in Japan," previously cited; also EIA, Japan country overview.
20. The 1.73 trillion yen figure is cited in *Tokyo Enerugi Foramu,* Tokyo, April 1998. Paik in *Oil and Gas in Northeast Asia,* previously cited, uses a figure of $31 billion through fiscal 1990, which would not be inconsistent with the $41 billion figure. I have not been able to obtain documentation, but three in separate ministries and/or state-funded companies, senior Japanese sources confirmed the figure to the author.
21. Foreign Broadcast Information Service, June 29, 1998. Quoted in Tokyo Enerugi Foramu, previously cited. April 1998.
22. See Alexandra Harney, "Japan Urged to Change Oil Policy," *Financial Times,* May 18, 1999.
23. IEA, World Energy Outlook, 1998; also EIA International Energy Outlook 1999.
24. Ministry of Foreign Affairs, Tokyo website, December 1999.
25. See Yasuhiko Wada, "Japan's Long-term E &P Investment Policies in the Asia Pacific," paper presented at the 8th Centre for Global Energy Studies Conference, London, April 1998.
26. See Amy Jaffe, presentation at "Wast Asian Energy Markets and Energy Cooperation in Northeast Asia," Baker Institute workshop, December 14, 1999. Generally, transport costs for pipelines over 3,000 km tend to be 2–3 times more than LNG.
27. See "Gas Pipeline Network Key to Outlook for Japan's Energy Mix, N.E. Asian Gas Grid," *Oil & Gas Journal,* July 13, 1998, pp. 23–28.
28. "Japan: Country Overview," EIA, Washington, May 1999, is the source of statistics in this paragraph.
29. See "Energy in Japan," MITI , Tokyo, 1999, previously cited.
30. See Jon Choy, "Serious Accident Strengthens Doubts about Japan's Nuclear Power Industry," Japan Economic Institute Report, October 8, 1999.
31. See Nuclear Fuel, January 13, 1996.
32. Japanese nuclear industry officials in background interviews with author, June and October 1999.
33. Mainichi Shimbun, August 1, 1994. For a detailed, well-documented analysis of Japan's consideration of nuclear weapons see Selig S. Harrison, "Japan and Nuclear Weapons," in *Japan's Nuclear Future,* ed. Harrison, Carnegie Endowment, Washington, D.C., 1996.
34. See for example, Howard W. French, "Accident Makes Japan Re-examine A-Plants," *New York Times,* January 13, 2000; also Peter Landers, "Uranium

Leak Feeds Anti-Nuclear Sentiment in Japan, *Wall Street Journal*, October 4, 1999.
35. Japanese nuclear industry official, background interview with author, December 1999.
36. These calculations were done by Amy Meyers Jaffe of the Baker Institute, December 1999.
37. "Energy Strategies and Military Strategies in Asia," report prepared for the Office of Net Assessments, Department of Defense, September 1999.
38. Cited in *Oil & Gas Journal,* January 19, 1998.
39. "South Korea: Country Overview," EIA, Washington, D.C., November 1999; 1999 estimates, ROK Ministry of Finance, author's calculations.
40. "APEC Energy Demand and Supply Outlook," Asia-Pacific Energy Research Centre, Tokyo, 1998.
41. Own calculations from BP Amoco Statistical Review of World Energy.
42. See Paul Hornswell, previously cited, chapter 5.
43. Paik, previously cited, and own calculations from BP Amoco Statistical Review of World Energy.
44. Presentation by Hyun-joon Chang, Korean Energy Economics Institute (KEEI), December 1999; author's own calculations.
45. South Korea: Country Overview, EIA, November 1999, previously cited.
46. For details of South Korean efforts to develop Russian natural gas, see Keun-Wook Paik, previously cited, especially Chapter 6.
47. Unpublished presentation by Hyun-Joon Chang, Korean Energy Economics Institute, December 1999
48. South Korea: Country Brief, EIA, November 1999.
49. KEEI forecasts growth at over 5 percent, but that assumed a longer recovery period from the 1997 financial crisis. The APEC energy research center (APERC), in its business-as-usual scenario, forecasts 11 percent growth in gas demand to 2010.
50. Korean Ministry of Finance official, interview with author, June 1999; KEEI researcher in discussion with author, December 1999.
51. South Korea: Country Brief, EIA, previously cited.
52. See Mo Tai-joon, "Mechanical Failure Blamed for Radiation Leak," *Chosen Ilbo,* Seoul, October 6, 1999.
53. See Foreign Broadcast Information Service (FBIS), Special Memorandum, "South Korea's Emerging Nuclear Potential," February 22, 1996.
54. For the most detailed assessment of the North Korean (Democratic People's Republic of Korea, or DPRK) energy sector see David F. Von Hippel and Peter Hayes, "DPRK Energy Sector: Current Status and Scenarios for 2000 and 2005," paper for an IIE conference, "Economic Integration of the Korean Peninsula," September 1997.
55. For a comprehensive, if pessimistic critique of the DPRK economy, see Marcus Noland, "The North Korean Economy," in *Joint U.S.-Korean Acad-*

emic Studies, vol. 6, 1996, Korean Economic Institute (KEI), pp. 127–178. Noland's analysis suggests sweeping reform under the current regime is highly improbable. See also Noland "The Korean Economy," Institute for International Economics, 2000, for forecasts on the cost of reconstruction.

56. "Energy Strategies and Military Strategies in Asia," unclassified report for Advisor to the Secretary of Defense for Net Assessment, September 1999, previously cited.

Chapter 8: Southeast Asia: Tigers as Regional Energy Link

1. The terms "Southeast Asia" and "ASEAN" are used here interchangeably, as ASEAN encompasses the ten states stretching from Burma to the Philippines, I include Papua New Guinea also as it is an energy producer. It is unclear if or when East Timor would join ASEAN.
2. See Henry J. Kenny, "An Analysis of Possible Threats to Shipping in Key Southeast Asian Sea Lanes," Center for Naval Analysis, Alexandria, Virginia, February 1996.
3. See World Bank, "The East Asian Miracle," Washington, D.C., 1994.
4. For purposes of analysis, Australia a non-ASEAN nation, is not included here. But it is a significant oil, gas, and coal supplier to East Asia, with 45 trillion ft. of proved gas reserves, 3 billion barrels of proved oil reserves (640,000 b/pd production), and 9 percent of world coal proven reserves, according to BP Amoco Statistical Review of World Energy 1999.
5. BP Amoco Statistical Review of World Energy, 1999, IEA, World Energy Outlook, 1998, author estimates.
6. BP Amoco Statistical Review of World Energy, 1999, also EIA, International Energy Outlook, 1999.
7. See Paul Hornswell, *Oil in Asia,* London, Oxford University Press, 1997, particularly chapter 10.
8. Ibid.
9. Paul Griggs, V.P., Strategic Planning BHP Petroleum, Ltd., Presentation at the 8th CGES Annual Conference, London, April 28, 1998.
10. See "Oil, Peace Would Play Parts in East Timor Independence," *Washington Times,* August 20, 1999; also "Timor Gap ZOC Now Void, Countries to Negotiate 16 Finds," *Oil & Gas Journal,* November 1, 1999 (Internet edition).
11. "Oiling the Wheels," *Asiaweek,* October 24, 1997.
12. BP Amoco Statistical Review of World Energy, 1999.
13. "Indonesia: Country Brief," Energy Information Administration, Washington, D.C., February 2000.
14. BP Amoco Statistical Review of World Energy, 1999.
15. See Budi Sudarsono, "Energy as a Development Resource: The Indonesian Experience," in Shankar Sharma and Fereidun Fesharaki, (eds.) *Energy Mar-*

ket and Policies in ASEAN, Institute of Southeast Asian Studies, Singapore, 1991.
16. "Indonesia Considers Legislation that would End Pertamina's 30 Year Petroleum Monopoly," Oil & Gas Journal, July 26, 1999 (Internet edition).
17. Ibid.
18. See John McBeth, "Slippery Road," Far Eastern Economic Review, December 24, 1998.
19. See Ronald Ripple, "The Maturing of Southeast Asia's Natural Gas Sector and the Growing Role of Gas in the Power Sector," in Natural Gas in Asia: Facts and Fiction, PECC Energy Forum Washington, D.C., 1998
20. See Daniel Yergin, The Prize, Simon and Schuster, New York, 1991, for a fuller historical account of the Singapore oil story.
21. See Paul Hornswell, Oil in Asia, chapter 6, for a detailed analysis of Singapore's oil sector.
22. See Richard Borsuk, "Singapore Barrels Ahead on Jurong Island," Asian Wall Street Journal, Weekly Edition, May 31–June 1, 1999.
23. Oil and gas figures from BP Amoco Statistical Review of World Energy, 1999.
24. "Malaysia: Country Brief," EIA, Washington, D.C., May 1999, and author's calculations.
25. EIA, previously cited, and author's estimate.
26. See "Malay-Thai JDA Gas Infrastructure Project Poised for Launch," Oil & Gas Journal, September 27, 1999 (Internet edition).
27. See S. Jayasinakaran, "Savior Complex," Far Eastern Economic Review, August 12, 1999, for an analysis of Petronas.
28. BP Amoco Statistical Review of World Energy, 1999.
29. EIA, International Energy Outlook, 1999.
30. See "Goodnight Vietnam," The Economist, January 8, 2000, for an analysis of the hiatus in Vietnam's reforms.
31. EIA, "Thailand Country Brief," January 1999, and "Philippines Country Brief," July 1999, BP Amoco, Statistical Review of World energy, author's calculations.
32. Philippines Country Analysis Brief, U.S. Energy Information Agency, April 2000.
33. See "Natural Gas in Asia: Facts and Fiction," PECC Energy Forum, Washington, D.C., 1998, for an exceptionally lucid explanation of the prospects and obstacles to realizing the development of natural gas and multilateral gas cooperative ventures.
34. See "Recommendations Concerning Accelerating Investment in Natural Gas Supplies, Infrastructure and Trading Networks in the APEC Region," Initiative of the APEC Energy Ministers, October 1998.
35. See "Oil Piracy Poses Growing Menace to Tanker Traffic in South China Sea," Oil & Gas Journal, October 18, 1999 (Internet edition).

36. See Henry J. Kenny, "An Analysis of Possible Threats to Shipping in Key Southeast Asian Sea Lanes," Center for Naval Analysis, Alexandria, Virginia, 1996.
37. *The Military Balance,* 1998 International Institute of Strategic Studies, London.

Chapter 9: Energy and Asian Security: Fueling Conflict or Cooperation?

1. See Humphrey Hawksley and Simon Holberton, *Dragon Strike: A Novel of the Coming War with China,* New York, St. Martin's Press, 1999.
2. Kent E. Calder, *Asia's Deadly Triangle,* London, Nicholas Brealey Publishing, 1997, p. 200. Also published in the United States as *Pacific Defense,* New York, William Morrow, 1996.
3. See Alastair Iain Johnston, "China's Militarized Interstate Dispute Behavior, 1049–92: A First Cut at the Data," *The China Quarterly,* no. 153, March 1998.
4. See Evan A. Feigenbaum, "China's Military Posture and the New Economic Geopolitics," *Survival,* Summer 1999, pp. 71–88.
5. See Michael Richardson, "Indonesia Plans War Games to Caution China," *International Herald Tribune,* August 16, 1996.
6. "Energy Strategies and Military Strategies in Asia," prepared for the Advisor to the Secretary of Defense for Net Assessment, September 1999.
7. See for example, Mark J. Valencia, "Energy an Insecurity in Asia," Autumn 1997, for the "energy scarcity" perspective.
8. *The Petroleum Economist;* July 1995
9. "Oil Prospective Spratlys Still a Flashpoint," *Oil & Gas Journal,* October 25, 1999 (Internet edition).
10. See Fereidun Fesharaki, "Oil Markets an Energy Security in Northeast Asia," in University of California, Institute on Global Conflict and Cooperation; Policy Paper #35, edited by Michael Stankiewicz, February 1998.
11. Author interviews with U.S. oil executives, June 1999, October 1999, November 1999, January 2000.
12. Figures cited in Michael C. Lynch, "The Nature of Energy Security," in the M.I.T. Japan Program Report, September/October 1997.
13. See Nicholas D. Kristof, "Asian Tensions Rise Over Seas' Wealth, *New York Times,* May 19, 1996.
14. See Robert K. Massie, *Dreadnought: Britain, Germany, and the Coming of the Great War,* Balantine Books paper edition 1992. This is a remarkable account of the naval arms race leading to World War I.
15. See Michael D. Bordo, Barry Eichengreen, and Douglas A. Irwin, "Is Globalization Today Really Different than Globalization a Hundred Years Ago?," paper for the Brookings Institute Trade Forum, April 15, 1999.

16. Thomas Paine, *Collected Writings,* vol. 1, London, 1894.
17. J. S. Mill, cited in Donald Kagan, *The Origins of War,* New York, Doubleday, 1995, p. 2.
18. See John Keegan, *The First World War,* New York, Alfred A. Knopf, 1998, pp. 10–12.
19. See China's 1998 Defense White Paper for its critique of the U.S. alliance system.
20. Background interview with author, Beijing, June 1999.
21. See Ji Guoxing, "SLOC Security in the Asia-Pacific," Asia-Pacific Center for Security Studies, Honolulu, 2000.
22. See John H. Noer, *Chokepoints: Maritime Concerns in Southeast Asia.* Washington, D.C.: National Defense University Press (in cooperation with the Center for Naval Analysis), 1996.
23. John H. Noer, *Chokepoints,* p. 4.
24. Noer, previously cited, p. 52.
25. See Al Troner, "Japan and the Russian Far East: The Economics and Competitive Impact of Least Cost Gas Imports," Working Paper for Japanese energy Security study, Baker Institute for Public Policy, Rice University, May 2000.
26. See Nayan Chanda, "Foot in the Water," *Far East Economic Review,* March 9, 2000.
27. See Robert A. Manning, "PACATOM: Nuclear Cooperation in Asia," *The Washington Quarterly,* Spring 1997, vol. 20, no. 2.

Bibliography

Adelman, M. A. *The Genie Out of the Bottle: World Oil Since 1970.* Cambridge: MIT Press, 1995.

Agency of Natural Resources and Energy (MITI), "Enerugi: Mirai kara no Keishou," [Warning Bells from the Future]. Tokyo, 1996.

Alternatives to Traditional Transportation Fuels, 1996. Washington, D.C.: Energy Information Administration, 1997.

APEC Energy Demand and Supply Outlook. Tokyo: Asia Pacific Energy Research Centre, 1998.

Armitage, Richard. "U.S.-Japan Relations in the Middle East." John Colabrese, ed., *The United States, Japan, and the Middle East.* Washington, D.C.: The Middle East Institute, 1997.

Atomic Energy Commission of Japan. "Long-Term Program for Research, Development and Utilization of Nuclear Energy." Tokyo, 1994.

Battelle Memorial Institute. "China's Electric Power Options: An Analysis of Economic and Environmental Costs." Advanced International Studies United, June 1998.

Bhattacharyaa, A. "Energy Conservation in Petroleum Industry," in *Strategy for Conservation in India*, ed. Pradeep Chaturuedi. New Delhi: Concept Publishing Company, 1996.

Blanchard, Odile. "Energy Consumption and Modes of Industrialization: Four Developing Countries." *Energy Policy* 20 (December 1992).

Brzezinski, Zbigniew. *The Grand Chessboard: American Primacy and Its Geostrategic Imperatives.* New York: Basic Books, 1997.

Cambridge Energy Research Associates. *Energy and the Environment: The New Landscape of Public Opinion.* Cambridge: Cambridge Energy Research Associates, 1990.

Carpenter, William M., and David G. Wiencek, eds. *Asian Security Handbook: An Assessment of Political-Security Issues in the Asia-Pacific Region*. Armonk, New York: M. E. Sharpe, 1996.

"Central Asia should serve as Eurasian 'buffer zone.'" *Yomiuri Shimbun*, 5 December 1997.

Christoffersen, Gaye. "China's Intentions for Russian and Central Asian Oil and Gas." *National Bureau of Asian Research Analysis* 9:2 (March 1998).

China's Worldwide Quest for Energy Security, International Energy Agency, Paris: IEA Publications, 2000.

Clawson, Patrick L., ed. *Energy and National Security in the 21ˢᵗ Century*. Washington, D.C.: National Defense University Press, 1995.

Cordesman, Anthony. *Are Energy Wars Still Possible?* Washington, D.C.: Center for Strategic and International Studies, 1999.

———. "The United States, Japan, and the Gulf: Meeting External Challenges." in *The United States, Japan, and the Middle East*, ed. John Calbrese. Washington, D.C.: The Middle East Institute, 1997.

Cossa, A. Ralph. *PACATOM: Building Confidence and Enhancing Nuclear Transparency*. Honolulu: Pacific Forum CSIS, 1998.

———. *The U.S.–DPRK Agreed Framework: Is it Still Viable? Is it Enough?* Honolulu: Pacific Forum CSIS, 1999.

"East Asia: a new market for Russian gas." *Petroleum Economist* (September 1996).

Eberstadt, Nicholas. *Korea Approaches Unification*. Armonk, NY: M. E. Sharpe, 1995.

Ehrlich, Paul R. and Anne H. Ehrlich. *The Population Explosion*. New York: Simon & Schuster, 1990.

Ellings, Richard J. and Sheldon W. Simon, eds. *Southeast Asian Security in the New Millennium*. Armonk, New York: M. E. Sharpe, 1996.

Engler, Robert. *The Brotherhood of Oil: Energy Policy and the Public Interest*. Chicago: University of Chicago Press, 1977.

"Far East: Natural Gas Leads Current, Future Construction." *Pipe Line and Gas Industry* 11:80 (November 1997).

Feis, Herbert. *Petroleum and American Foreign Policy*. Stanford: Food Research Institute, 1944.

———. *The Road to Pearl Harbor: The Coming of War Between the United States and Japan*. Princeton: Princeton University Press, 1950.

Gore, Al. *Earth in the Balance: Ecology and the Human Spirit*. Boston: Houghton Mifflin, 1992.

Grubb, Michael, et al. *Emerging Energy Technologies: Impacts and Policy Implications*. Dartmouth, Great Britain: Royal Institute of International Affairs, 1992.

Han, Yong-Sup. "Korea's Security Strategy for 21ˢᵗ Century: Cooperation and Conflict." *Korea Focus* 5:4 (July-August 1997).

Harrison, Selig S. *China, Oil, and Asia: Conflict Ahead?* New York: Columbia University Press, 1977.

———. *Japan's Nuclear Future: the Plutonium Debate and East Asian Security*. Washington, D.C.: Brookings Institution, 1996.

Hiro, Dilip. *Between Marx and Muhammad: The Changing Face of Central Asia*. London: Harper Collins, 1994.

Hopkirk, Peter. *The Great Game: The Struggle for Empire in Central Asia*. New York: Kodansha International, 1992.

Horsnell, Paul. *Oil in Asia: Markets, Trading, Refining and Deregulation*. Oxford: Oxford University Press, 1997.

Hawksley, Humphrey and Simon Holberton. *Dragon Strike: A Novel of the Coming War with China*. New York: St. Martin's Press, 1999.

India's Need for Strategic Balance. New Delhi: Delhi Policy Group, 1996.

Kapstein, Ethan B. *The Insecure Alliance: Energy Crises and Western Politics Since 1944*. Oxford: Oxford University Press, 1990.

Kemp, Geoffrey. *Energy Superbowl: Strategic Politics and the Persian Gulf and Caspian Basin*. Washington, D.C.: Nixon Center for Peace and Freedom, 1997.

Kemp, Geoffrey and Robert E. Harkavy. *Strategic Geography and the Changing Middle East*. Washington, D.C.: Brookings Institution Press, 1997.

Kenny, Henry J. *An Analysis of Possible Threats to Shipping in Key Southeast Asian Sea Lanes*. Alexandria, VA: Center for Naval Analyses, 1996.

Kux, Dennis. *India and the United States: Estranged Democracies*. Fort McNair: National Defense University Press, 1993.

Leifer, Michael. *The ASEAN Regional Forum: Extending ASEAN's Model of Regional Security*. Oxford: Oxford University Press, 1996.

Mak, J. N. "The ASEAN Naval Build-up: Implications for the Regional Order." *Pacific Review* 8:2 (1995).

Manning, Robert A., Don Oberdorfer and Kathryn Weathersby. "Korean Questions." *Wilson Quarterly* (Summer 1999).

Manning, Robert A. and James J. Przystup. "Asia's Transition Diplomacy: Hedging Against Futureshock." *Survival* (Autumn 1999).

Mansourov, Alexandre Y. "The Origins, Evolution and Future of the North Korean Nuclear Program." *Korea and World Affairs* 19:1 (Spring 1995).

Massie, Robert K., *Dreadnought: Britain, Germany, and the Coming of the Great War*. New York: Random House, 1991.

May, Michael. "Energy and Security in East Asia." Asia-Pacific Research Center, Stanford University, January 1998.

Mazarr, Michael. *North Korea and the Bomb*. New York: St. Martins, 1995.

McRae, Hamish. *The World in 2020: Power, Culture and Prosperity*. Boston: Harvard Business School Press, 1994.

Menon, Rajan. "Treacherous Terrain: The Political and Security Dimensions of Energy Development in the Caspian Sea Zone." *National Bureau of Asian Research Analysis* 9:1 (February 1998).

Mitchell, John V et al. *The New Geopolitics of Energy*. London: Royal Institute of International Affairs, 1996.

238   Bibliography

Natural Gas in Asia: Facts and Fiction. Washington, D.C.: PECC Energy Forum, 1998.

Noer, John H. and David Gregory. *Chokepoints: Maritime Economic Concerns in Southeast Asia.* Washington, D.C.: National Defense University Press, 1996.

Nuclear Power Generation and Fuel Cycle Report, 1996. Washington, D.C.: Energy Information Administration, 1996.

Pearce, Joan, ed. *The Third Oil shock: The Effects of Lower Oil Prices.* London: Royal Institute of International Affairs, 1983.

Petroleum 1996: Issues and Trends. Washington, D.C.: Energy Information Administration, 1997.

"Possible Gas Line to India." *Petroleum Economist* (June 1989).

Renewable Energy Annual 1997, Volume 1. Washington, D.C.: Energy Information Administration, 1998.

Rohwer, Jim. *Asia Rising: How History's Biggest Middle Class Will Change the World.* Butterworth-Heinemann Asia, 1995.

Roy-Chaudhury, Rahul. "An Energy Security Policy for India: The Case for Oil and Natural Gas," on www.idsa-india.org, January 1998.

Rubin, Barry. "China's Middle East Strategy." *Middle East Review of International Affairs* 3:1 (March 1999).

"Russian Crisis Spills Over." *Petroleum Economist* 65:11 (November 1998).

Salamaeh, Mahmoud G. "China, Oil and the Risk of Regional Conflict." *Survival* 37:4 (Winter 1995–96).

Sampson, Anthony. *The Seven Sisters: The Great Oil Companies and the World They Created.* Rev. ed. London: Coronet, 1988.

Samuels, Richard J. *The Business of the Japanese State: Energy Markets in Comparative Historical Perspective.* Ithaca: Cornell University Press, 1987.

Schwartz, Peter, Peter Leyden, and Joel Hyatt. *The Long Boom: A Vision for the Coming Age of Prosperity.* New York: Perseus Books, 1999.

Short-Term Energy Outlook, Quarterly Projections. Washington, D.C.: Energy Information Administration, 1998 (First Quarter).

Simon, Julian L. *The Ultimate Resource 2.* Princeton: Princeton University Press, 1996.

Singh, Ranjeet J., ed. *Investigating Confidence-Building Measures in the Asia-Pacific Region.* Washington, D.C.: The Henry L. Stimson Center, 1999. Report No. 28.

Snyder, Jed C., ed. *After Empire: The Emerging Geopolitics of Central Asia.* Washington, D.C.: National Defense University Press, 1995.

Sokolsky, Richard and Tanya Charlick-Paley. *NATO and Caspian Security: A Mission Too Far?* Santa Monica, CA: RAND, 1999.

Statistical Review of World Energy 1999. London, England: BP Amoco, 1999.

Stephan, John. *The Russian Far East: A History.* Stanford University Press, 1994.

Tanham, George K. *Indian Strategic Thought: An Interpretive Essay.* Santa Monica: RAND, 1993.

Terzian, Philip. *OPEC: The Inside Story.* Trans. Michael Pallis. London: Zed Books, 1985.

Valencia, Mark. *China and the South China Sea Disputes*. Oxford: Oxford University Press, 1995.
Valencia, Mark and James P. Dorian. "Multilateral Cooperation in Northeast Asia's Energy Sector: Possibilities and Problems." University of California Institute on Global Conflict and Cooperation (February 1998).
van Creveld, Martin. *Supplying War: Logistics from Wallenstein to Patton*. Cambridge: Cambridge University Press, 1977.
Walker, Martin. "China and the New Era of Resource Scarcity." *World Policy Journal* (Spring 1996).
Whiting, Allen S. *China Eyes Japan*. Berkeley: University of California Press, 1989.
Winchester, Simon. *Pacific Rising: The Emergence of a New World Culture*. New York: Prentice Hall Press, 1991.
World Energy Outlook. Paris: International Energy Agency, 1998.
Yager, Joseph A. and Shelley M. Matsuba. *The Energy Balance in Northeast Asia*. Washington, D.C.: Brookings Institution, 1984.
Yajima, Masayuki. *Deregulatory Reforms of the Electricity Supply Industry*. Westport, CT: Quorum Books, 1997.
Yergin, Daniel. *The Prize*. New York: Simon & Schuster, 1992.

Index

Adelman, M. A., 24
Agency for Natural Resources and Energy (ANRE), 149
AIOC (Azerbaijan International Operating Company), 48, 50, 51, 53
Akins, James, 18
Albright, Madeline, 42, 52
Aliev, Haydar, 51
Anderson, Roger N., 31
Angell, Norman, 197
ANRE (Agency for Natural Resources and Energy), 149
APEC (Asia-Pacific Economic Cooperation), 183–84, 196–97, 199, 204, 205
ARCO, 95, 107, 154, 176
ARF (ASEAN Regional Forum), 197, 199, 204, 205
ASEAN (Association of Southeast Asian Nations), 68, 103, 134, 154, 169–72, 180, 182, 183–86, 188, 189, 190, 199, 200, 204, 205, 206
ASEAN Power Grid, 184, 203
ASEAN Regional Forum (ARF), 197, 199, 204, 205
ASEAN+3, 197, 199
Asia, 14, 29, 59–84, 187–207
  energy consumption, 20, 67–71
  energy futures, 63–66
  energy security, 59–60, 187–207
  geopolitics, 59–84
  supply and demand, 71–74
Asia-Pacific Economic Cooperation (APEC), 183–84, 196–97, 199, 204, 205
Asia's Deadly Triangle, 187
"Asian Miracle," 2, 179
Asian territories, 191–93
Association of Southeast Asian Nations (ASEAN), 68, 103. *See also* ASEAN
Atlantic Monthly, 20
automobiles, 8, 19, 34–37, *102*, 104–5
Azerbaijan International Operating Company (AIOC), 48, 50, 51, 53

Bernabe, Franco, 23
Beyond the Limits, 9
Bharatiya Janata Party (BJP), 121–22, 126, 127
Bijur, Peter, 34
BJP (Bharatiya Janata Party), 121–22, 126, 127
BP Amoco, 34, 38, 45, 46, 50
BP Amoco Statistical Review of World Energy, 23
Bracken, Paul, 197
Brown, Lester, 7, 15, 20
Brzezinski, Zbigniew, 43

Index 241

Calder, Kent, 79, 187
Campbell, Dr. Colin J., 22, 23, 24, 26, 27–28
Carson, Rachel, 3
Caspian Sea, 41–48, *49*, 50–58
Center for Naval Analysis (CNA), 185, 201, 202
Central Asia, 41, 42–43, 47, 58
China, 2, 6, 15, 33, 43, 48, 58, 85–117
  economic reform, 87–93, 94, 107–9
  energy policy, 85–117
  natural gas consumption, *129*
  nuclear power, 98–100
  oil consumption, 20, 86, 103–4
  oil industry, 105–17
  petroleum imports, *86*
  power sources, 96–100
  supply and demand, 94–96, 105–10
China National Chemicals Import and Export Corporation (CNOOC), 105, 106, 107, 109–10
Chinese National Petroleum Company (CNPC), 86–87, 91. *See also* CNPC
Chokepoints, 201
Chrysler, 19, 35, 36
Churchill, Winston, 4
Clinton, Bill, 42, 52
Club of Rome, 7, 8, 9, 10, 20
CNA (Center for Naval Analysis), 201, 202
CNOOC (China National Chemicals Import and Export Corporation), 105, 106, 107, 109–10
CNPC (Chinese National Petroleum Company), 86–87, 91, 105, 106, 107, 108, 109–10
Cohen, Joel, 5
"Coming Conflict with China, The," 2
"Coming War with Japan, The," 2
"Commission on Japan's Goals in the Twenty-first Century," 147

Daimler-Benz, 35
de Condorcet, Marquis, 5

Dragon Strike, 187, 190
Dreadnought, 139, 194
"Drifting Towards Disaster," 20
Dutta, Sujit, 137–38

Earth in the Balance, 3
East Asia, 20, 29. *See also* Asia
"East Asian Miracle, The," 2
Easterbrook, Gregg, 4
economics. *See* geoeconomics
"Economics of Exhaustible Resources, The," 24
Economist, 15, 43, 182
EEZs (Exclusive Economic Zones), 185, 189
Ehrlich, Paul, 3, 7, 20
EIA (Energy Information Agency), 66, 67, 72, 101, 130
energy conflicts, 187–207
energy consumption, 6, 68–71
  India, *125*
  Japan, *146*
  South Korea, *146*
  world, *88*
Energy Information Agency (EIA), 66, 67, 72. *See also* EIA
energy initiatives, 202–7
energy policies, 1, 137–42, 187–207
  China, 85–117
  India, 119–42
  Japan, 143–67
  Korea, 143–67
  South Korea, 161
  Southeast Asia, 169–86
energy production, *73*
  China, 100, *101*, 102–14
  Southeast Asia, 171–74
energy security, 20, 79
  Asia, 59–60, 187–207
  China, 87–88, 101, 107
  India, 137–42
  Japan, 145, 147
  policies, 137–42, 187–207
energy, supply of, 17–39
Enron Corporation, 122, 130–31, 164
"Essay on the Principle of Population," 6

Exclusive Economic Zones (EEZs), 185, 189
Exxon, 78, 154, 182
Exxon Mobil, 38

Famine 1975, 7
Ferdinand, Archduke, 194
Forbes, 21, 23
Ford Motor Company, 19, 35
Foreign Affairs, 18, 21
Foreign Ministry, 149
Foreign Policy, 54
Forrester, Dr. Jay W., 7–8
Friedberg, Aaron, 197

GAIL (Gas Authority of India Limited), 129, 131, 136
Gandhi, Indira, 121
Gandhi, Rajiv, 121
gas. *See* natural gas
Gas Authority of India Limited (GAIL), 129, 131, 136
gas, liquid, 32–33. *See also* liquefied natural gas (LNG)
GDP (gross domestic product), 61, *64*, *69*, 193
   China, 68, 90
   India, 120, 125
   Indonesia, 65
   Japan, 65, 145, 147
   Southeast Asia, 171
   United States, 68
General Motors (GM), 19, 34, 35
geoeconomics, 6, 17, 67, 77–84, 114–15
geopolitics, 41, 50–53, 54–58, 59–84, 114, 133–37, 193–95, 197, 204
Global 2000 Report to the President, The, 10
Glomar Explorer, 32
GM (General Motors), 19, 34, 35
Gopalakrishnan, Adinarayana, 141–42
Gore, Al, 3, 4, 42, 107
Gorlov, Dr. Alexander, 36
"Great Games," 41, 42, 107
Great Illusion, The, 197

gross domestic product (GDP), 61, *64*, *69*. *See also* GDP

Habibie, B. J., 174
Hashemi, Mehdi, 134
Hashimoto, Ryutaro, 160, 192
Hassina, Sheikh, 137
He, Admiral Zhang, 80
Hee, Park Chung, 162
Hitler, Adolf, 5
Honda, 35
Horiuchi, Mitsuo, 152
Hotelling argument, 24
Hotelling, Harold, 24
How Many People Can the Earth Support?, 5
Hubbert bell curve, 24, 32
Hubbert, M. King, 24
Huntington, Samuel, 115
Hussein, Saddam, 83
Hydrocarbon Vision 2025, 127
Hyundai, 163

IAEA (International Atomic Energy Agency), 158, 206
IEA (International Energy Agency), 17, 19, 21, 23, 37, 100, 101, 102, 116, 148, 153, 171, 205
Il, Kim Jong, 166
ILSA (Iran-Libya Sanctions Act), 52
independent power producers (IPPs), 96, 155, 164
India, 6, 42–43
   economic reform, 126–28, 133–37
   energy challenges, 122–26
   energy consumption, *125*
   energy policy, 119–42
   geopolitics, 133–37
   market solutions, 126–28
   Middle East connection, 131–33
   natural gas consumption, *125*, 128, *129*, 130–31
   oil consumption, 20, *125*, 128–31
   similarities to China, 119–21

Index

India's Nuclear Bomb, 140
Indonesia, 5, 65, 174–76
International Atomic Energy Agency (IAEA), 158, 206
International Energy Agency (IEA), 17, 19, 21. *See also* IEA
investment trust companies (ITICs), 90
investments, 30, 87–88, 92, 109–10
IPPs (independent power producers), 96, 155, 164
Iran-Libya Sanctions Act (ILSA), 52
Issues in Science and Technology, 21
ITICs (investment trust companies), 90

JAEC (Japan Atomic Energy Commission), 156–57
Japan, 2, 5, 14, 19, 58
  energy consumption, *146*
  energy needs, 145–49, 159–61
  energy policy, 143–67
  natural gas consumption, *129*, 153–56
  nuclear energy, 156–59
  oil consumption, 20, 153–56
  oil dilemmas, 149–53
  similarities to Korea, 144–45
Japan Atomic Energy Commission (JAEC), 156–57
Japan China Oil Development Corporation, 150–51
Japan National Oil Company (JNOC), 107, 150–52, 154
Japanese Export-Import Bank, 126
JDA (Joint Development Area) of Malaysia-Thailand, 180, 183
JNOC (Japan National Oil Company), 107, 150–52, 154
Jung, Kim Dae, 65, 165–66

Kant, Immanuel, 197
Kemp, Geoffrey, 83
Khatemi, Mohamed, 52–53
Kissinger, Henry, 54, 148
Korea, 20, 143–67. *See also* North Korea; South Korea

Kyoto Protocol on Global Warming, 145, 156

Laherrere, Jean H., 22
Law of the Sea (LOS) Treaty, 185
life expectancy, *12*
Light Water Reactors (LWRs), 157, 158, 165–66
Limits to Growth, 7, 8, 9, 20
liquefied natural gas (LNG), 32–33, 36, 72, 78. *See also* LNG
LNG (liquefied natural gas), 72, 78, 112, 113, 137
London Observer, 21
LOS (Law of the Sea) Treaty, 185
Lugar, Richard, 21
Lukman, Rilwanu, 77
LWRs (Light Water Reactors), 157, 158, 165–66
Lynch, Michael C., 24, 26, 27–28

Mainichi Shimbun, 158
Malaysia, 179–81
Malaysia-Thailand Joint Development Area (JDA), 180, 183
Malthus, Thomas Robert, 5–6
Malthusian predications, 3, 6–7, 15
Massie, Robert K., 139
May, Dr. Michael, 38
Middle East, 22, 28, 29, 58, 74–75, *76*, 77, 131–33
military conflict, 187–90, 199–200, 202
military involvement, 6, 54–55, 79–81, 82–83, 139–40, 199
Mill, John Stuart, 197
Minister of Mines and Energy, 177
Ministry of Land and Natural Resources, 106
Ministry of Petroleum and Natural Gas, 129
Ministry of Trade and Industry (MITI), 106. *See also* MITI
Ministry of Transport, 149
Missile Technology Control Regime (MTCR), 82

MITI (Ministry of Trade and Industry), 106, 147, 149, 151, 152, 155, 156
Mitsubishi Motors, 35, 154
Miyazawa Initiative, 154
Mobil, 182
Models of Doom, 8
Monroe Doctrine, 141
MTCR (Missile Technology Control Regime), 82

Nakayama, Taro, 154
Narayanan, K. R., 127
National People's Congress (NPC), 90, 94, 97, 108, 189
NATO, 43, 54–57, 204
natural gas, *49*, 68
  China, 71–72, 93, 95, 110–17
  consumption of, *129*
  Japan, 153–56
  pipelines, 111, 112, *113*
NELP (New Exploration Licensing Policy), 130, 132
neo-Malthusian predictions, 1, 14, 188
New Exploration Licensing Policy (NELP), 130, 132
New Republic, 4
New York Times, 41, 42, 47, 192
Newsweek, 35
non-OPEC production, 25, *26*, 31, 147
Non-Proliferation Treaty (NPT), 82, 98, 195, 198
Noroil, 26
North Korea, 2, 15, 145, 165–66. *See also* Korea
NPC (National People's Congress), 90, 94, 97, 108, 189
NPT (Non-Proliferation Treaty), 82, 98, 195, 198
nuclear accidents, 100, 157–58, 159, 165
nuclear energy, 36, 71, 82, *195*
  China, 93, 97, 98–100
  East Asia, 195–202
  Japan, 156–59
  South Korea, 164–67

nuclear weapons, 82, 83, 99, 140–41, 165–66, 195–96

Obuchi, Prime Minister, 147
ODA (Official Development Assistance), 154
OECD, 95, 101, 156, 163, 179, 180, 205
OECD countries, 11, 14, 59, 142, 145
Office of Net Assessments, 139, 159, 190
Official Development Assistance (ODA), 154
Official Secrets Act, 141
oil, 17–39, *76*, *94*
  consumption of, 19, 20, 25, 29, 62, *68*, 86, *88*, 103, 153–56
  drilling problems, 45–46
  investment trends, 30, 87, 92, 109–10
  world demand, 17, 67, 71–74, 94–96, 105–10
Oil & Gas Journal, 155
Oil and Natural Gas Corporation (ONGC), 129, 130, 136
"Oil Crisis: This Time the Wolf is Here," 18
Oil India Limited (OIL), 129, 130
oil industry, 9, 17–39, 105–17
oil markets, 17, 19, 20, 28–29, 30, 39, 44, 64–65, 75
oil nexus, 74–77
OIL (Oil India Limited), 129, 130
oil predictions, 17, 19–20, 21, 22–25, *26*, *27*
oil prices, *18*, 19–20, 28–29, 33
oil production, 21–23, 25–29, 30–32, 62, 73, 95–96, 100–104, 114–17
oil reserves, 20, 23, 24–26, 44, 72–74, *94*
  Caspian Sea, 43, 45, 50, 57
  Central Asia, 41–58
  Middle East, 41–58
  political barriers, 50–53, 54–58
oil shocks, 18, 19, 20–21, 75

Index 245

oil transport routes, 47, 49, 81, 170. See also pipelines
Olcott, Martha, 54
ONGC (Oil and Natural Gas Corporation), 129, 130, 136
OPEC (Organization of Petroleum Exporting Countries), 17–18, 19, 23, 25, 29, 77–79, 116, 148–49, 161, 174, 191
Organization of Petroleum Exporting Countries (OPEC), 17–18. See also OPEC
Our Stolen Future, 4

Paine, Thomas, 197
Parade, 21, 41
Parliamentary Committee on Energy, 123
Partnership for Peace (PFP), 54
Peng, Li, 95, 97, 104, 107
Pentagon, 139, 159, 160, 190
People's Daily, 192
Perkovich, George, 140
Persian Gulf, 25, 29, 30, 41, 44, 48, 58
Petroleum Administration Bureaus, 106
Petroleum Authority of Thailand (PTT), 183
Petroleum Ministry, 105
Petronas, 78, 180–81
Petronas Towers, 169, 179
PFP (Partnership for Peace), 54
Philippines, 182–83
pipelines, 42, 46, 47–48, 49, 50–53, 81, 92–93, 103, 111, 112, 113, 114, 170, 184, 203–4
politics, world, 41, 50–53, 54–58, 59–84. See also geopolitics
Population Bomb, The, 7
population growth, 1, 3, 5, 6, 7, 10–12, 13, 14–15
Potential Implications of Trends in World Population, Food Production and Climate, 9
Power Vision 2010, 127
Principles of Political Economy, 197

PTT (Petroleum Authority of Thailand), 183

reactors, 99–100, 157–58
Richardson, Bill, 42, 47
Rongji, Premier Zhu, 90
Royal/Dutch Shell, 78
Rushdie, Salman, 52

SAARC (South Asian Association for Regional Cooperation), 134
Sakhalin projects, 204
Sam, Kim Young, 163
Science and Technology Agency (STA), 149
Scientific American, 21, 22
sealanes, 81, 116, 205
Sealanes of Communication (SLOCs), 138, 139, 159. See also SLOCs
SEBs (State Electricity Boards), 124–25, 127
Sen, Amartya, 15
SETC (State Economic and Trade Commission), 106, 108
Shell Oil, 24
Silent Spring, 3
Simon, Julian, 9, 16, 25
Singh, Jaswant, 138
Sinochem, 105, 106
Sinopec, 105, 106, 107, 109
SLOCs (Sealanes of Communication), 138, 139, 159, 197, 201–2, 205
SOEs (state-owned enterprises), 89, 90, 91, 108, 109
South Asian Association for Regional Cooperation (SAARC), 134
South Korea, 2, 143–67. See also Korea
energy consumption, 146
energy policy, 161
natural gas consumption, 163–64
nuclear energy, 164–67
oil strategies, 162–63
similarities to Japan, 144–45, 161–62
Southeast Asia, 169–86
energy production, 170, 171, 172, 173–74
oil reserves, 172

246 Index

STA (Science and Technology Agency), 149
Stalin, Joseph, 15
Starr, S. Frederick, 54
State Economic and Trade Commission (SETC), 106, 108
State Electricity Boards (SEBs), 124–25, 127
State of the World Report, 15
state-owned enterprises (SOEs), 89, 90, 91, 108, 109
Steinfeld, Edward, 89
Suharto, President, 174, 176–78

Ta Kung Pao, 115
Talbot, Strobe, 43
Tanaka, Kakuei, 148
technological revolution, 28, 30–32, 34–37, 74, 173
Texaco, 33, 34, 95
Thailand, 182–83
This Moment on Earth, 4
Three Gorges Dam, 97, 98
Three Mile Island, 100, 157–58
Time, 35
Timor Gap Treaty, 174, 203
Toyota, 35

UN Fund for Population Activities, 12
UN Population Conference, 14
UN Security Council, 138, 205
United Nations, 10, 12, 14
United States, 3, 5, 11, 42, 43
  oil consumption, 20, 29
  oil reserves, 25, 30, 31
U.S. Department of Energy, 21, 67
U.S. Department of State, 43
U.S. Department of the Interior, 17
U.S. Energy Information Agency, 19–20
U.S. European Command, 54
U.S. EXIM bank, 98

U.S. Geological Survey (USGS), 17, 22, 23
U.S. Navy, 31, 58, 116, 138, 197, 201, 202
U.S. Security and Exchange Commission, 23
USGS (U.S. Geological Survey), 17, 22, 23
USGS World Petroleum, Assessment 2000, 22

Vajpayee, Atali Bihari, 122
vehicles, 8, 19, 34–37, *102*, 104–5
Vietnam, 181–82

Wada, Yasuhiko, 154
Wahid, Aburrahman, 175, 177
Wall Street Journal, 43
Warsaw Pact, 48
Woolsey, R. James, 21
World Bank, 96, 110, 126
World Health Organization, 90–91, 126
world politics, 41, 50–53, 54–58, 59–84. *See also* geopolitics
"World Population Prospects," 14
World Trade Organization (WTO), 91, 109, 117, 198
Worldwatch Institute, 7, 12
WTO (World Trade Organization), 91, 109, 117, 198

Xiaopeng, Deng, 59–60
Xu, Xiaoje, 107

Yamani, Ahmed Zaki, 77
Yeltsin, Boris, 163
Yergin, Daniel, 60
Yung, Chung Ju, 163

Zemin, Jiang, 74, 77–78